FRACTIONAL ORDER SYSTEMS
Modeling and Control Applications

WORLD SCIENTIFIC SERIES ON NONLINEAR SCIENCE

Editor: Leon O. Chua
University of California, Berkeley

*To view the complete list of the published volumes in the series, please visit:
http://www.worldscibooks.com/series/wssnsa_series.shtml

WORLD SCIENTIFIC SERIES ON NONLINEAR SCIENCE Series A Vol. 72

Series Editor: Leon O. Chua

FRACTIONAL ORDER SYSTEMS

Modeling and Control Applications

Riccardo Caponetto
Giovanni Dongola
Luigi Fortuna
University of Catania, Italy

Ivo Petráš
Technical University of Košice, Slovakia

World Scientific

NEW JERSEY · LONDON · SINGAPORE · BEIJING · SHANGHAI · HONG KONG · TAIPEI · CHENNAI

Chapter three is on fractional order chaotic systems. In this chapter, a survey of well-known chaotic systems is presented. Mathematical models of nonlinear dynamical systems contain the fractional derivatives. Total order of the system is less than three, however, the chaotical phenomena, as for example, in strange attractors can be observed in such systems.

In chapter four the operator s^m, where m is a real number, is approximated via the binomial expansion of the backward difference and then a hardware implementation of differintegral operator is proposed using Field Programmable Gate Array (FPGA). This building block represents the basic element to implement fractional order control systems.

Chapter five is devoted to microprocessor implementation of the fractional order controllers. Fundamentals on discrete approximations of a fractional operator as well as control algorithm for implementation of the controllers are described. Also presented are three examples of the discrete fractional order controllers implemented on PIC, PC with PCL card, and PLC, respectively. A real measurement and obtained results are shown for each particular case. Some concluding remarks close this chapter.

Chapter six is dedicated to the implementation of the fractional order PID controller by using the analog counter part of FPGA that is Field Programmable Analog Array (FPAA).

Chapter seven presents a possible implementation of an Integrated Circuit by using the switched capacitor technology. The aim of the chapter is to start a research activity that can provide an integrated circuit implementing differintegral operators.

Chapter eight concludes this book showing an useful modelling application of fractional order system on Ionic Polymeric Metal Composite (IMPC) membranes. Going beyond the IMPC, the proposed modelling approach shows that it is possible to obtain low order fractional order models instead of bigger order integer one.

More than 140 references are listed and cited in the book, even if it cannot be a complete bibliography for this area of interest. Readers can find many other references related to this topic.

Riccardo Caponetto
Giovanni Dongola
Luigi Fortuna
Ivo Petráš

Preface

This book is devoted to fractional order systems, their applications to modelling and control. It is based on derivatives and integrals of arbitrary (real) order, fractional differential equations and methods of their solution, approximations and implementation techniques.

The advantages of fractional calculus have been described and pointed out in the last few decades by many authors. It has been shown that the fractional order models of real systems are regularly more adequate than usually used integer order models.

Applications of these fractional order models are in many fields, as for example, rheology, mechanics, chemistry, physics, bioengineering, robotics and many others.

At the same time, fractional integrals and derivatives are also applied to the theory of control of dynamical systems, when the controlled system and/or the controller is described by fractional differential equations.

The main goal of the book is to present applications and implementations of fractional order systems. It provides only a brief theoretical introduction to fractional order system dedicating almost all the space to the modelling issue, fractional chaotic system control and fractional order controller theory and realization.

The book is suitable for advanced undergraduates and graduate students.

It is organized as follows:

Chapter one is a brief introduction to the fractional order systems. Some historical notes, definitions and fundamentals are described.

Chapter two is dedicated to Fractional Order PID Controller defining their stability regions when first order with time delay plant have to be controlled in closed loop.

Chapter three is on fractional order chaotic systems. In this chapter, a survey of well-known chaotic systems is presented. Mathematical models of nonlinear dynamical systems contain the fractional derivatives. Total order of the system is less than three, however, the chaotical phenomena, as for example, in strange attractors can be observed in such systems.

In chapter four the operator s^m, where m is a real number, is approximated via the binomial expansion of the backward difference and then a hardware implementation of differintegral operator is proposed using Field Programmable Gate Array (FPGA). This building block represents the basic element to implement fractional order control systems.

Chapter five is devoted to microprocessor implementation of the fractional order controllers. Fundamentals on discrete approximations of a fractional operator as well as control algorithm for implementation of the controllers are described. Also presented are three examples of the discrete fractional order controllers implemented on PIC, PC with PCL card, and PLC, respectively. A real measurement and obtained results are shown for each particular case. Some concluding remarks close this chapter.

Chapter six is dedicated to the implementation of the fractional order PID controller by using the analog counter part of FPGA that is Field Programmable Analog Array (FPAA).

Chapter seven presents a possible implementation of an Integrated Circuit by using the switched capacitor technology. The aim of the chapter is to start a research activity that can provide an integrated circuit implementing differintegral operators.

Chapter eight concludes this book showing an useful modelling application of fractional order system on Ionic Polymeric Metal Composite (IMPC) membranes. Going beyond the IMPC, the proposed modelling approach shows that it is possible to obtain low order fractional order models instead of bigger order integer one.

More than 140 references are listed and cited in the book, even if it cannot be a complete bibliography for this area of interest. Readers can find many other references related to this topic.

Riccardo Caponetto
Giovanni Dongola
Luigi Fortuna
Ivo Petráš

Acknowledgments

There are several people to whom the authors are obliged for their help and support.

Ivo Petráš (Technical University of Košice, Slovakia) would like to express his thanks to Prof. Igor Podlubny, Prof. Ján Terpák, Prof. Ľubomir Dorčák, and Prof. Imrich Koštial (Technical University of Košice, Slovakia), Prof. Paul O'Leary (Montanuniversitat of Leoben, Austria), Prof. YangQuan Chen (Utah State University in Logan, USA), and Prof. Blas M. Vinagre (University of Extremadura in Badajoz, Spain) for their help, exchange of information and the fruitful discussions. The author would also like to thank a number of colleagues and friends who supported his work. Last but not least, Ivo Petráš is also thankful to his entire family for their understanding and support.

Particular thanks from Giovanni Dongola, Luigi Fortuna and Riccardo Caponetto to Professor Leon Chua from University of California, Berkeley, for his continuous encouragement in their research life.

Contents

List of Figures

xiii

List of Tables

Chapter 1

Fractional Order Systems

In this chapter a brief introduction to Fractional Calculus will be given. Starting from the historical roots of the Science of Fractional Calculus, this chapter aims to define fundamental notions and observable behaviours of FOS.

1.1 Fractional Order Differintegral Operator: Historical Notes

The concept of the differentiation operator $D = \frac{d}{dx}$ is familiar to all those who have studied elementary calculus. For suitable functions, the n-th derivative of f, namely $D^n f(x) = \frac{d^n f(x)}{dx^n}$, is well defined when n is a positive integer. In 1695 L'Hopital asked Leibniz what meaning could be ascribed to $D^n f$ if n were a fraction. Since that time fractional calculus has drawn the attention of many famous mathematicians, such as Euler, Laplace, Fourier, Abel, Liouville, Riemann, and Laurent. But it was not until 1884 that the theory of generalized operators achieved such a level in its development so as to make it suitable as a point of departure for the modern mathematician. By then the theory had been extended to include D^m operators, where m could be rational or irrational, positive or negative, real or complex. Thus the name *fractional calculus* became somewhat of a misnomer. A better description might be *differentiation and integration to an arbitrary order*. However, we shall adhere to tradition and refer to this theory as fractional calculus.

During the investigations of the general theory and applications of differintegrals (a term that was coined to avoid the cumbersome alternate "derivatives or integrals to arbitrary order"), it was discovered that, while this subject is old, dating back at least to Leibniz in its theory and to

Heaviside in its application, it has been studied relatively little since the early papers which only hinted at its scope. In the last several years a mild revival of interest in the subject seems to have taken place, but the application of these ideas has not yet been fully exposed, primarily because of their unfamiliarity. Our studies have convinced us that differintegral operators may be applied advantageously in many diverse areas. Within mathematics, the subject is in contact with a very large segment of classical analysis and provides a unifying theme for a great number of well known, and some new, results. Applications outside mathematics include otherwise unrelated topics such as: transmission line theory, chemical analysis of aqueous solutions, design of heat-flux meters, rheology of soils, growth of intergranular grooves on metal surfaces, quantum mechanical calculations, and dissemination of atmospheric pollutants.

Fractional derivatives provide an excellent instrument for the description of memory and hereditary properties of various materials and processes. This is the main advantage of fractional derivatives in comparison with classical integer-order models, in which such effects are in fact neglected. The advantages of fractional derivatives become apparent in modeling mechanical and electrical properties of real materials, as well as in the description of rheological properties of rocks, and in many other fields.

Fractional integrals and derivatives also appear in the theory of control of dynamical systems, when the controlled system or/and the controller is described by a fractional differential equation. The mathematical modeling and simulation of systems and processes, based on the description of their properties in terms of fractional derivatives, naturally leads to differential equations of fractional order and to the necessity to solve such equations.

The idea of fractional derivatives and integrals seems to be quite a strange topic, very hard to explain, due to the fact that, unlike commonly used differential operators, it is not related to some important geometrical meaning, such as the trend of functions or their convexity. For this reason, this mathematical tool could be judged "far from reality". But many physical phenomena have "intrinsic" fractional order description and so fractional order calculus is necessary in order to explain them.

1.2   Preliminaries and Definitions

Fractional systems, or more non integer order systems, can be considered as a generalization of integer order systems [Oldham (2006)], [Ross (1975)], [Sabatier (2007)], [Kilbas (2006)] and [Das (2007)].

Fractional calculus is a generalization of integration and differentiation to non-integer order fundamental operator $_aD_t^r$, where a and t are the limits of the operation and $r \in \mathbb{R}$. The continuous integro-differential operator is defined as

$$_aD_t^r = \begin{cases} \frac{d^r}{dt^r} & : r > 0, \\ 1 & : r = 0, \\ \int_a^t (d\tau)^{-r} & : r < 0. \end{cases}$$

The three equivalent definitions most frequently used for the general fractional differintegral are the Grünwald-Letnikov (GL) definition, the Riemann-Liouville (RL) and the Caputo definition [Oldham (2006)], [Podlubny (1999a)].

The GL definition is given by

$$_aD_t^r f(t) = \lim_{h \to 0} h^{-r} \sum_{j=0}^{[\frac{t-a}{h}]} (-1)^j \binom{r}{j} f(t - jh), \tag{1.1}$$

where [.] means the integer part.

The RL definition is given as

$$_aD_t^r f(t) = \frac{1}{\Gamma(n-r)} \frac{d^n}{dt^n} \int_a^t \frac{f(\tau)}{(t-\tau)^{r-n+1}} d\tau, \tag{1.2}$$

for $(n - 1 < r < n)$ and where $\Gamma(.)$ is the *Gamma* function.

The Caputo definition can be written as

$$_aD_t^r f(t) = \frac{1}{\Gamma(r-n)} \int_a^t \frac{f^{(n)}(\tau)}{(t-\tau)^{r-n+1}} d\tau, \tag{1.3}$$

for $(n-1 < r < n)$. The initial conditions for the fractional order differential equations with the Caputo derivatives are in the same form as for the integer-order differential equations.

In the above definition, $\Gamma(m)$ is the factorial function, defined for positive real m, by the following expression:

$$\Gamma(m) = \int_0^\infty e^{-u} u^{m-1} du \tag{1.4}$$

for which, when m is an integer, it holds that:

$$\Gamma(m + 1) = m! \tag{1.5}$$

The definition of fractional derivative easily derives by taking an n order derivative (n suitable integer) of a m order integral (m suitable non integer) to obtain an $n - m = q$ order one:

$$\frac{d^q f(t)}{dt^q} = \frac{d^{n-m} f(t)}{dt^{n-m}} = \frac{1}{\Gamma(m)} \frac{d^n}{dt^n} \int_0^t (t-y)^{m-1} f(y) dy \tag{1.6}$$

It must be noted that for $q = 1$ ($n = 2, m = 1$), (1.6) becomes the canonical first order derivative.

Furthermore, most of the derivation rules holding for integer order derivatives can be extended to the non integer order case. However, we must remember the following exceptions in order to avoid frequent mistakes:

Leibniz rule:

$$\frac{d^m\big(f(t)g(t)\big)}{dt^m} \neq \frac{d^m f(t)}{dt^m}g(t) + \frac{d^m g(t)}{dt^m}f(t) \tag{1.7}$$

Chain rule:

$$\frac{d^m f\big(g(t)\big)}{dt^m} \neq \frac{d^m f}{dg^m}\frac{d^m g(t)}{dt^m} \tag{1.8}$$

Equivalent rules for fractional differentiation do exist, but since they are not so immediate (they involve series expansions and therefore infinite terms have to be considered), their use is limited to few cases.

1.3 Laplace Transforms and System Representation

In system theory the analysis of dynamical behaviours is often made by means of transfer functions. With this in view, the introduction of the Laplace transform of non integer order derivatives is necessary for an optimal study. Fortunately, not very big differences can be found with respect to the classical case, confirming the utility of this mathematical tool even for fractional systems. Inverse Laplace transformation is also useful for time-domain representation of systems for which only the frequency response is known. The most general formula is the following [Oldham (2006)]:

$$L\left\{\frac{d^m f(t)}{dt^m}\right\} = s^m L\{f(t)\} - \sum_{k=0}^{n-1} s^k \left[\frac{d^{m-1-k}f(t)}{dt^{m-1-k}}\right]_{t=0} \tag{1.9}$$

where n is an integer such that $n - 1 < m < n$.

The above expression becomes very simple if all the derivatives are zero:

$$L\left\{\frac{d^m f(t)}{dt^m}\right\} = s^m L\{f(t)\} \tag{1.10}$$

Expression (1.10) is very useful in order to calculate the inverse Laplace transform of elementary transfer functions, such as non integer order integrators $1/s^m$. In fact, replacing m with $-m$ and considering $f(t) = \delta(t)$, the Dirac impulse, by means of the definition (1.3), it holds that:

$$L\left\{\frac{t^{m-1}}{\Gamma(m)}\right\} = \frac{1}{s^m}; \qquad L^{-1}\left\{\frac{1}{s^m}\right\} = \frac{t^{m-1}}{\Gamma(m)} \qquad (1.11)$$

that is the impulse response of a non integer order integrator.

Another important result can be derived by the well known frequency translation formula $L^{-1}\{F(s+a)\} = e^{-at}L^{-1}\{F(s)\}$ applied to (1.11):

$$L^{-1}\left\{\frac{1}{(s+a)^m}\right\} = \frac{t^{m-1}e^{-at}}{\Gamma(m)} \qquad (1.12)$$

Expression (1.12) is fundamental in fractional calculus for many reasons: firstly because it gives the impulse response of the generic fractional system $F(s) = k/(s+a)^m$, and secondly because it suggests a tool for deriving a time domain representation with a finite number of terms. In fact, traditional routes involved Taylor expansion of the given transfer function in order to obtain a differential equation of the type:

$$\sum_{k=0}^{\infty} a_k \frac{d^k y(t)}{dt^k} = u(t) \qquad (1.13)$$

where $u(t)$ is the generic input and $y(t)$ the output of the fractional system. For equation (1.13) the following expression of a_k can be calculated:

$$a_k = \frac{(-1)^k \Gamma(k-m)}{k!\Gamma(-m)a^{k-m}} = u(t) \qquad (1.14)$$

and, since

$$\lim_{k\to\infty} |a_k| = 0 \qquad (1.15)$$

there will exist a certain number N (usually very large) that allows the approximation:

$$\sum_{k=0}^{\infty} a_k \frac{d^k y(t)}{dt^k} \approx \sum_{k=0}^{N} a_k \frac{d^k y(t)}{dt^k} \qquad (1.16)$$

The alternative representation descending from the use of (1.10) and (1.12) leads to the following results:

$$\frac{Y(s)}{U(s)} = \frac{1}{(s+a)^m}; \qquad (s+a)^m Y(s) = U(s) \qquad (1.17)$$

Substituting s with $(s-a)$ on both sides of the previous equation, we obtain:

$$s^m Y(s-a) = U(s-a) \qquad (1.18)$$

and taking the inverse Laplace transform, it results that:

$$L^{-1}\{s^m Y(s-a)\} = L^{-1}\{U(s-a)\}; \qquad \frac{d^m}{dt^m}\left[e^{at}y(t)\right] = e^{at}u(t) \quad (1.19)$$

Now we have a non integer order differential equation, with a finite number of terms, having an exponential time dependence on both the input and the output of the system. A time-varying state space representation can be the following:

$$\begin{cases} \frac{d^m x(t)}{dt^m} = A(t)x(t) + B(t)u(t) \\ y(t) = C(t)x(t) + D(t)u(t) \end{cases} \quad (1.20)$$

where $A(t) = 0, B(t) = e^{at}, C(t) = e^{-at}, D(t) = 0$. To solve(1.19) and (1.20), expression (1.3) is commonly used, which is the general formula of non integer integration; however, only in few cases is its solution analytically derivable. With this in view, particular numerical routines are adopted, as will be discussed successively.

1.4   General Properties of the Fractional System

It is now useful to introduce the most important features of fractional systems. They will be discussed using the same tools usually adopted for integer order systems which allow an easy comparison among the two different behaviors. For example, let us focus on Bode Diagrams, that is the principal tool in systems and control theory. Considering $F(s) = k/(s+a)^m$ and assuming $s = j\omega$, we obtain:

$$F(j\omega) = \left[\frac{k^{1/m}}{(j\omega/p+1)}\right]^m = \left[\left|\frac{k^{1/m}}{(j\omega/p+1)}\right| e^{j\varphi\left[\frac{k^{1/m}}{(j\omega/p+1)}\right]}\right]^m$$

$$= \left|\frac{k^{1/m}}{(j\omega/p+1)}\right|^m e^{j\omega\varphi\left[\frac{k^{1/m}}{(j\omega/p+1)}\right]} \quad (1.21)$$

and, therefore the magnitude expressed in decibels is

$$|F(j\omega)|_{dB} = 20\log_{10}\left[\frac{k^{1/m}}{\sqrt{\omega^2/p^2+1}}\right]^m$$

$$= 20\log_{10} k - 20m\log_{10}\sqrt{\omega^2/p^2+1} \quad (1.22)$$

It must be noted that, if $\omega \to \infty$, (1.22) becomes $-20m\log_{10}(\omega/p)$ resulting, on a semi-logarithmic plane, in a line having slope $-20m\frac{dB}{dec}$ (instead of $-20dB/dec$ for first order systems). This fact is useful to plot an asymptotic diagram whose maximum error e_{max} can be found close to the pole $\omega = -p$. This error can be calculated as follows:

$$e_{max} = \left| |F(jp)|_{dB,app} - |F(jp)|_{dB} \right| = \left| -20\log_{10}\left(\frac{\omega}{p}\right)\right|_{\omega=p}$$

$$\left. -\left(-20m\log_{10}\sqrt{\frac{\omega^2}{p^2}+1}\right)\right|_{\omega=p} \cong 3mdB \qquad (1.23)$$

while for first order systems this value is $3dB$.

Examples of magnitude Bode diagrams are reported in Fig. 1.1. It is quite evident that the fractional order m modulates the slope of the magnitude diagram, providing a useful parameter for the open loop synthesis of the controller.

Regarding the phase displacement, it must be noted that, considering the exponent of expression (1.21), it holds that:

$$\varphi[F(j\omega)] = m\varphi\left[\frac{k^{1/m}}{(j\omega/p+1)}\right] = -m\arctan\frac{\omega}{p} \qquad (1.24)$$

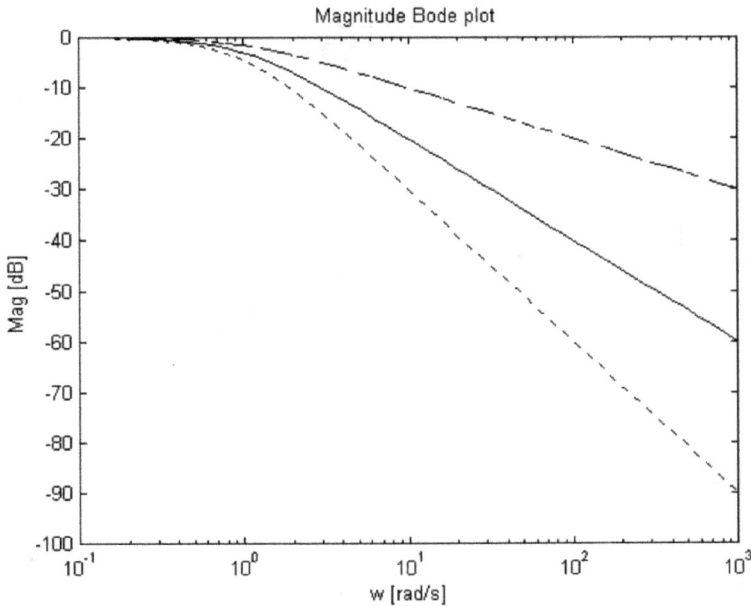

Fig. 1.1 Magnitude Bode Plot of fractional system $F(s) = 1/(s+1)^m$ with $m = 1$ (solid), $m = 0.5$ (dashed), $m = 1.5$ (dotted).

Expression (1.24) shows that m modulates the scale of the phase law. In fact, it can be easily seen that for $\omega \to \infty$ the phase angle approaches $-m\pi/2$ instead of $-\pi/2$ typical of first order systems. Examples of phase diagrams are depicted in Fig. 1.2.

Let us now consider the impulse canonical responses. From (1.12) the following expression for impulse response can be written:

$$f(t) = L^{-1}\left\{\frac{k}{(s/p+1)^m}\right\} = L^{-1}\left\{\frac{kp^m}{(s+p)^m}\right\} = kp^m \frac{t^{m-1}e^{-pt}}{\Gamma(m)} \quad (1.25)$$

The most relevant consideration regarding (1.25) is that, if $m < 1$, for $t \to 0$ $f(t)$ is infinite (See Fig. 1.3).

However it can be shown that, for any positive m, $f(t)$ satisfies the hypothesis of Cauchy's theorem on the existence of the integral [Kaplan (1992)], and therefore the step response can be calculated.

If we introduce the following definition of Incomplete Gamma Function:

$$\Gamma(m,x) = \int_0^x e^{-y}y^{m-1}dy \quad (1.26)$$

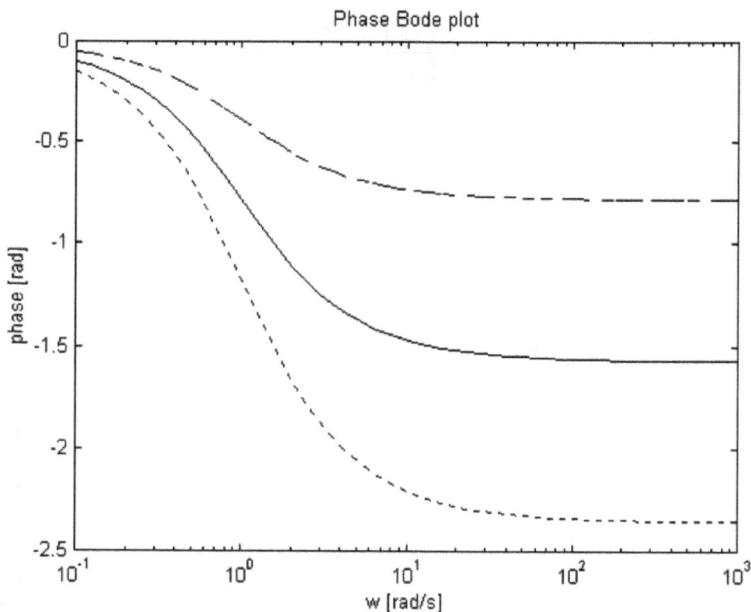

Fig. 1.2　Phase Bode Plot of fractional system $F(s) = 1/(s+1)^m$ with $m = 1$ (solid), $m = 0.5$ (dashed), $m = 1.5$ (dotted).

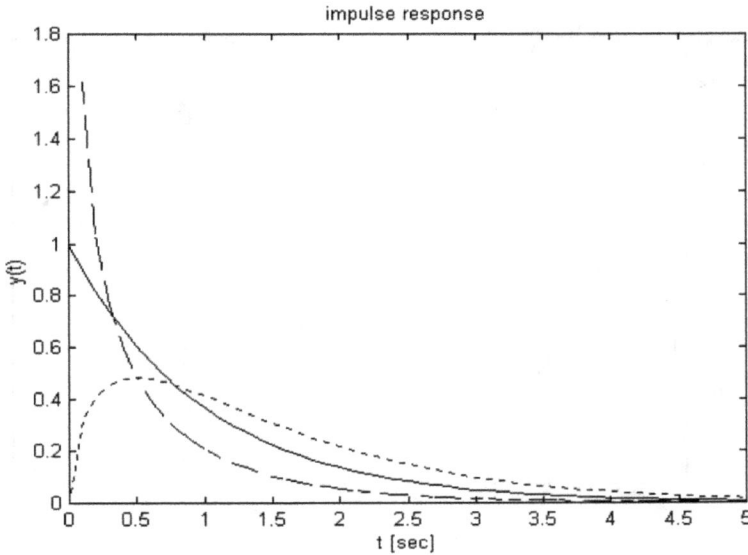

Fig. 1.3 Impulse response of a fractional system for different values of m (dashed), $m = 0.5$ (solid), $m = 1$ (dotted), $m = 1.5$.

the step response of our fractional system can be expressed by the simple formula:

$$g(t) = \int_0^t f(t)dt = k\frac{\Gamma(m, pt)}{\Gamma(m)} \tag{1.27}$$

If we look at Fig. 1.4, a faster response of the system having $m < 1$ can be observed. This fact, due to the infinite value of the derivative for $t = 0$, may be successfully exploited for all those applications in which a high speed is required; for example, in fast control schemes [Oustaloup (1983)] and image processing [Arena (1998)].

1.5 Impulse Response of a General Fractional System

In previous sections, the theory regarding fractional systems having the form $F(s) = k/(s + a)^m$ has been discussed. However, there are several examples available in literature that can be associated to the following transfer function:

$$F(s) = \frac{b}{s^m + a} \tag{1.28}$$

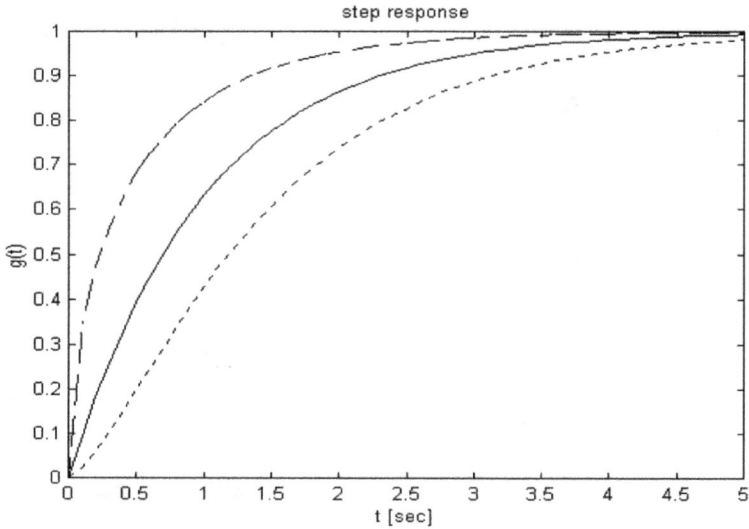

Fig. 1.4   Step response of a fractional system for different values of m (dashed), $m = 0.5$ (solid), $m = 1$ (dotted), $m = 1.5$.

This is quite a frequent case, especially in the time domain. In fact, if we consider $F(s) = X(s)/U(s)$, it can be easily shown that the corresponding differential equation is:

$$\frac{d^m x(t)}{dt^m} = -ax(t) + bu(t) \tag{1.29}$$

The inverse Laplace transform of (1.28) is not analytically derivable, but a series expansion of it can be easily obtained. Expanding (1.28) it results that:

$$F(s) = \frac{b}{s^m + a} = \frac{b}{s^m} \sum_{n=0}^{\infty} \frac{(-a)^n}{s^{nm}} \tag{1.30}$$

and, by using (1.11) for each term of the sum, the impulsive response is:

$$f(t) = L^{-1} \left[\frac{b}{s^m} \sum_{n=0}^{\infty} \frac{(-a)^n}{s^{nm}} \right] = bt^{m-1} \sum_{n=0}^{\infty} \frac{(-a)^n t^{nm}}{\Gamma(nm + m)} \tag{1.31}$$

whose graphs for different values of m and $a = b = 1$ are depicted in Fig. 1.5.

The Mittag-Leffler function:

$$E_m(x) = \sum_{n=0}^{\infty} \frac{x^n}{\Gamma(nm + 1)} \tag{1.32}$$

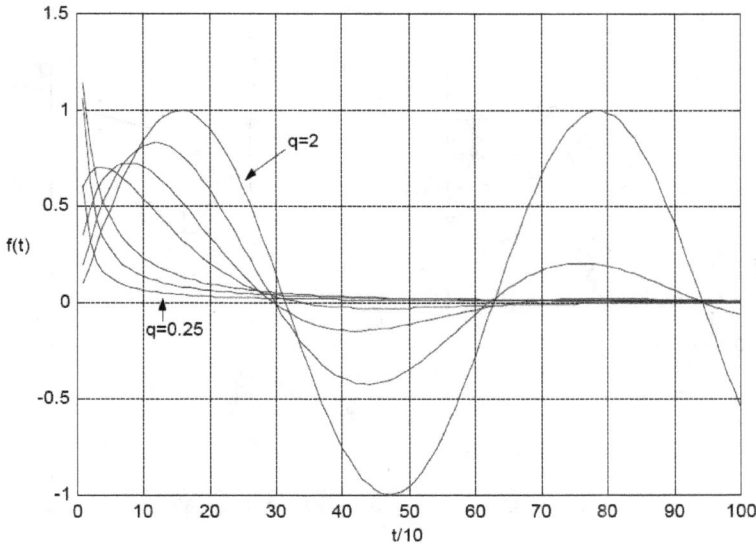

Fig. 1.5   The impulse response $f(t)$ as m varies from 0.25 to 2 in 0.25 increments.

is strictly related to (1.31) (see Fig. 1.6). In fact, it is possible to write it in terms of $E_m(.)$ as follows:

$$f(t) = \frac{d^{1-m} E_m(-at^m)}{dt^{1-m}} \qquad (1.33)$$

This fact is very useful in order to easily derive other canonical responses such as step response, ramp response, and so on.

For example, knowing that the Laplace transform of $E_m[-at^m]$ is:

$$L\{E_m[-at^m]\} = \frac{1}{s} \sum_{n=0}^{\infty} \left(\frac{-a}{s^m} \right)^n = \frac{1}{s} \left[\frac{s^m}{s^m + a} \right] \qquad (1.34)$$

the step response can be easily derived using (1.30) and (1.33). Assuming it to be $G(s)$ ($b = 1$ for simplicity), it can be written:

$$\frac{F(s)}{s} = \frac{1}{s} \left[\frac{1}{s^m + a} \right] = \frac{1/a}{s} \left[1 - \frac{s^m}{s^m + a} \right] \qquad (1.35)$$

and, therefore

$$g(t) = \frac{1}{a} \left[H(t) - E_m[-at^m] \right] = \frac{1/a}{s} \left[1 - \frac{s^m}{s^m + a} \right] \qquad (1.36)$$

where $H(t)$ is the Heaviside unit step function. The graph shown in Fig. 1.7, is not so different from that reported in Fig. 1.6.

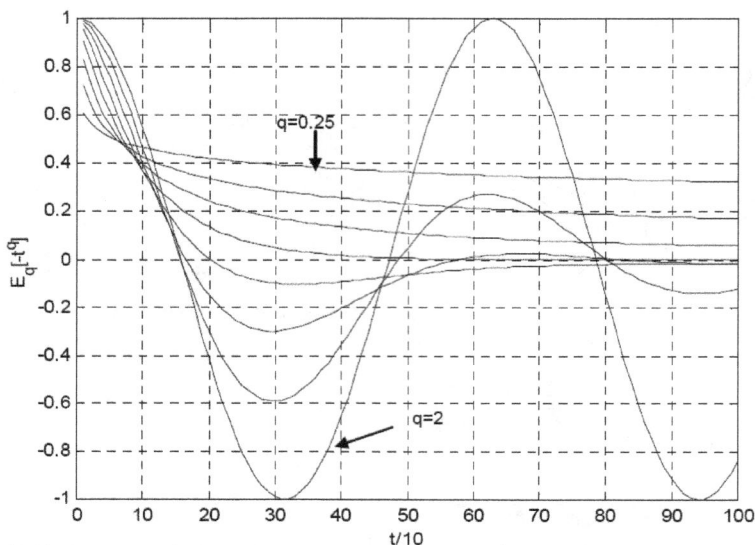

Fig. 1.6   The Mittag-Leffler function, $E_m[-t]$ as m varies from 0.25 to 2 in 0.25 increments.

1.6   Numerical Methods for Calculation of Fractional Derivatives and Integrals

For numerical calculation of fractional-order derivative we can use the relation (1.37) derived from the Grünwald-Letnikov definition (1.1). This approach is based on the fact that for a wide class of functions, three definitions - GL (1.1), RL (1.2), and Caputo's (1.3) - are equivalent. The relation for the explicit numerical approximation of r-th derivative at the points $kh, (k = 1, 2, \dots)$ has the following form [Podlubny (1999a)], [Vinagre (2003)], [Dorčák (1994)]:

$$_{(k-L_m/h)}D^r_{kh}f(t) \approx h^{-r} \sum_{j=0}^{k}(-1)^j \binom{r}{j} f_{k-j}, \qquad (1.37)$$

where L_m is the "memory length", h is the time step of the calculation and $(-1)^j \binom{r}{j}$ are binomial coefficients $c_j^{(r)}$, $(j = 0, 1, \dots)$. For their calculation we can use the following expression [Dorčák (1994)]:

$$c_0^{(r)} = 1, \qquad c_j^{(r)} = \left(1 - \frac{1+r}{j}\right) c_{j-1}^{(r)}. \qquad (1.38)$$

The described numerical method is so called the Power Series Expansion (PSE) of a generating function. It is important to note that PSE leads to

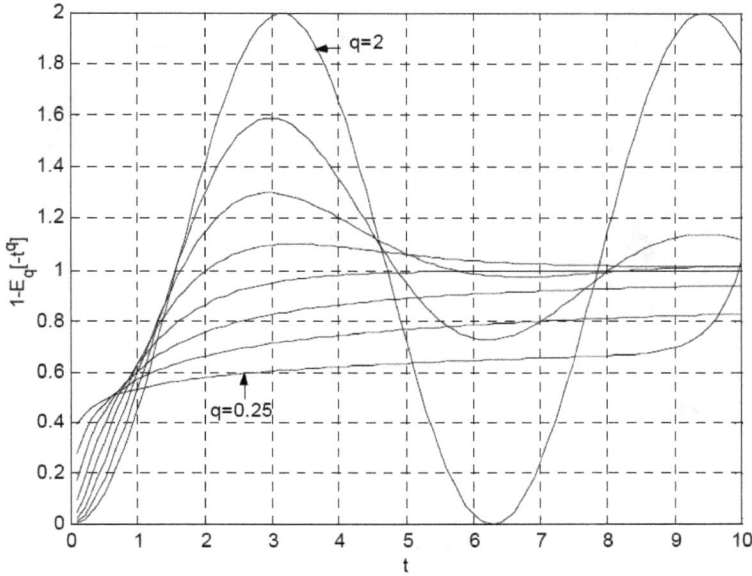

Fig. 1.7   The step response as m varies from 0.25 to 2 in 0.25 increments.

approximation in the form of polynomials, that is, the discretized fractional operator is in the form of a FIR filter, which has only zeros.

The resulting discrete transfer function, obtained by approximating fractional-order operators, can be expressed in the z-domain as:

$$
{}_0D_{kT}^{\pm r}\, G(z) = \frac{Y(z)}{F(z)} = \left(\frac{1}{T}\right)^{\pm r} \mathrm{PSE}\left\{\left(1 - z^{-1}\right)^{\pm r}\right\}_n \approx T^{\mp r} R_n(z^{-1}),
$$

(1.39)

where T is the sample period, $\mathrm{PSE}\{u\}$ denotes the function resulting from applying the power series expansion to the function u, $Y(z)$ is the Z transform of the output sequence $y(kT)$, $F(z)$ is the Z transform of the input sequence $f(kT)$, n is the order of the approximation, and R is polynomial of degree n, correspondingly, in the variable z^{-1}, $z = \exp(sT)$, and $k = 1, 2, \ldots$. Matlab routine dfod1() of this method can be downloaded from MathWorks, Inc. web (see [Petráš (2003a)]).

Another approach can be achieved by Continued Fraction Expansion (CFE) of the generating function and then the approximated fractional operator is in the form of a IIR filter, which has poles and zeros [Chen (2006b)].

By taking into account that our aim is to obtain equivalents to the fractional integrodifferential operators in the Laplace domain, $s^{\pm r}$, the result of such approximation for an irrational function, $G(s)$, can be expressed in the form:

$$G(s) \simeq a_0(s) + \cfrac{b_1(s)}{a_1(s) + \cfrac{b_2(s)}{a_2(s) + \cfrac{b_3(s)}{a_3(s) + \ldots}}}$$

$$= a_0(s) + \frac{b_1(s)}{a_1(s)+} \frac{b_2(s)}{a_2(s)+} \frac{b_3(s)}{a_3(s)+} \quad \ldots \tag{1.40}$$

where $a_i(s)$ and $b_i(s)$ are rational functions of the variable s, or are constants. The application of the method yields a rational function, which is an approximation of the irrational function $G(s)$.

In other words, for evaluation purposes, the rational approximations obtained by CFE frequently converge much more rapidly than the PSE and have a wider domain of convergence in the complex plane. On the other hand, the approximation by PSE and the short memory principle is convenient for the dynamical properties consideration.

For interpolation purposes, rational functions are sometimes superior to polynomials. This is, roughly speaking, due to their ability to model functions with poles. These techniques are based on the approximations of an irrational function, $G(s)$, by a rational function defined by the quotient of two polynomials in the variable s in frequency s-domain:

$$G(s) \simeq R_{i(i+1)\ldots(i+m)} = \frac{P_\mu(s)}{Q_\nu(s)} = \frac{p_0 + p_1 s + \ldots + p_\mu s^\mu}{q_0 + q_1 s + \ldots + q_\nu s^\nu}, \tag{1.41}$$

$$(m+1 = \mu + \nu + 1)$$

passing through the points $(s_i, G(s_i)), \ldots, (s_{i+m}, G(s_{i+m}))$.

The resulting discrete transfer function, obtained by approximating fractional-order operators, can be expressed as:

$$_0 D_{kT}^{\pm r} G(z) = \frac{Y(z)}{F(z)} = \left(\frac{2}{T}\right)^{\pm r} \text{CFE}\left\{\left(\frac{1-z^{-1}}{1+z^{-1}}\right)^{\pm r}\right\}_{p,n}$$

$$\approx \left(\frac{2}{T}\right)^{\pm r} \frac{P_p(z^{-1})}{Q_n(z^{-1})}, \tag{1.42}$$

where T is the sample period, $\text{CFE}\{u\}$ denotes the function resulting from applying the continued fraction expansion to the function u, $Y(z)$ is the Z transform of the output sequence $y(kT)$, $F(z)$ is the Z transform of the input sequence $f(kT)$, p and n are the orders of the approximation, and P and Q are polynomials of degrees p and n, correspondingly, in the variable

z^{-1}, and $k = 1, 2, \ldots$. Matlab routine dfod2() can be downloaded from MathWorks, Inc. web (see [Petráš (2003b)]).

For simulation purposes, here we present the Oustaloup recursive approximation (ORA) algorithm [Oustaloup (1983, 2008)]. The method is based on the approximation of a function of the form:

$$H(s) = s^r, \qquad r \in \mathrm{R}, \qquad r \in [-1; 1] \tag{1.43}$$

for the frequency range selected as (ω_b, ω_h) by a rational function:

$$\widehat{H}(s) = C_o \prod_{k=-N}^{N} \frac{s + \omega'_k}{s + \omega_k} \tag{1.44}$$

using the following set of synthesis formulas for zeros, poles and the gain:

$$\omega'_k = \omega_b \left(\frac{\omega_h}{\omega_b} \right)^{\frac{k+N+0.5(1-r)}{2N+1}},$$

$$\omega_k = \omega_b \left(\frac{\omega_h}{\omega_b} \right)^{\frac{k+N+0.5(1-r)}{2N+1}}, \tag{1.45}$$

$$C_o = \left(\frac{\omega_h}{\omega_b} \right)^{-\frac{r}{2}} \prod_{k=-N}^{N} \frac{\omega_k}{\omega'_k},$$

where ω_h, ω_b are the high and low transitional frequencies. An implementation of this algorithm in Matlab as a function ora_foc() is given in [Chen (2003)].

A detailed review of the various approximation methods and techniques (Carlson's [Carlson (1964)], Chareff's [Charef (2006)], CFE [Chen (2006b)], Oustaloup's [Oustaloup (1995)], etc.) for continuous and discrete fractional-order models in form of IIR and FIR filters was done in [Vinagre (2003)]. Some other approaches were described in [Podlubny (2002a)]. Last but not least we should mention the approach proposed by Hwang, which is based on the B-splines function [Hwnag (2002)] and Podlubny's matrix approach [Podlubny (2002b, 2009)].

The frequency domain approximation methods are not always reliable, especially in detecting chaos behaviour in nonlinear systems [Tavazoei (2007a, 2008b, 2007b)]. As has been shown, due to the error of approximation, numerical simulation may result in wrong conclusions, e.g. fake chaos is produced due to the implementation of the frequency domain approximation methods [Tavazoei (2007b)]. Simulation of the fractional-order system using the time domain methods is complicated and due to long memory characteristics of these systems requires a very long simulation time but on the other hand it is more accurate. Applying some ideas such as, for instance, the short memory principle [Podlubny (1999a)], we can reduce the computational cost of the time-domain methods. Results obtained by these methods are more reliable than those determined by using the frequency based approximation [Tavazoei (2008b)].

For numerical simulation of the fractional order system a method on the basis of the Adams-Bashforth-Moulton type predictor-corrector scheme has also been proposed [Deng (2007a)]. It is suitable for Caputo derivative because it just requires the initial conditions and for unknown function it has clear physical meaning. The method is based on the fact that fractional differential equation

$$D_t^q y(t) = f(y(t), t), \ \ y^{(k)}(0) = y_0^{(k)}, \ k = 0, 1, \ldots, m - 1$$

is equivalent to the Volterra integral equation

$$y(t) = \sum_{k=0}^{[q]-1} y_0^{(k)} \frac{t^k}{k!} + \frac{1}{\Gamma(q)} \int_0^t (t - \tau)^{q-1} f(\tau, y(\tau)) d\tau. \tag{1.46}$$

Discretizing the Volterra equation (1.46) for $t_n = nh$ $(n = 0, 1, \ldots, N)$, $h = T_{sim}/N$ and using the short memory principle (fixed or logarithmic [Ford (2001)]) we obtain a good numerical approximation of the true solution of a fractional differential equation while preserving the order of accuracy. A slightly improved predictor-corrector approach to solve the Fokker-Planck equation has been noted in [Deng (2007b)]. A collection of the various numerical algorithms was also presented in [Diethelm (2005)].

1.7   Fractional LTI Systems

A general fractional-order system can be described by a fractional differential equation of the form

$$a_n D^{\alpha_n} y(t) + a_{n-1} D^{\alpha_{n-1}} y(t) + \ldots + a_0 D^{\alpha_0} y(t)$$
$$= b_m D^{\beta_m} u(t) + b_{m-1} D^{\beta_{m-1}} u(t) + \ldots + b_0 D^{\beta_0} u(t), \tag{1.47}$$

where $D^\gamma \equiv {}_0 D_t^\gamma$ denotes the Riemann-Liouville or Caputo fractional derivative [Podlubny (1999a)], or by the corresponding transfer function of *incommensurate* real orders of the following form [Podlubny (1999a)]:

$$G(s) = \frac{b_m s^{\beta_m} + \ldots + b_1 s^{\beta_1} + b_0 s^{\beta_0}}{a_n s^{\alpha_n} + \ldots + a_1 s^{\alpha_1} + a_0 s^{\alpha_0}} = \frac{Q(s^{\beta_k})}{P(s^{\alpha_k})}, \tag{1.48}$$

where a_k $(k = 0, \ldots n)$, b_k $(k = 0, \ldots m)$ are constant, and α_k $(k = 0, \ldots n)$, β_k $(k = 0, \ldots m)$ are arbitrary real or rational numbers and without loss of generality they can be arranged as $\alpha_n > \alpha_{n-1} > \ldots > \alpha_0$, and $\beta_m > \beta_{m-1} > \ldots > \beta_0$.

The incommensurate order system (1.48) can also be expressed in commensurate form by the multivalued transfer function [Bayat (2008)]

$$H(s) = \frac{b_m s^{m/v} + \cdots + b_1 s^{1/v} + b_0}{a_n s^{n/v} + \cdots + a_1 s^{1/v} + a_0}, \quad (v > 1). \tag{1.49}$$

Note that every fractional order system can be expressed in the form (1.49) and domain of the $H(s)$ definition is a Riemann surface with v Riemann sheets [Lepage (1961)].

In the particular case of *commensurate* order systems, it holds that, $\alpha_k = \alpha k, \beta_k = \alpha k, (0 < \alpha < 1), \forall k \in Z$, and the transfer function has the following form:

$$G(s) = K_0 \frac{\sum_{k=0}^{M} b_k (s^\alpha)^k}{\sum_{k=0}^{N} a_k (s^\alpha)^k} = K_0 \frac{Q(s^\alpha)}{P(s^\alpha)} \tag{1.50}$$

With $N > M$, the function $G(s)$ becomes a proper rational function in the complex variable s^α which can be expanded in partial fractions of the following form:

$$G(s) = K_0 \left[\sum_{i=1}^{N} \frac{A_i}{s^\alpha + \lambda_i} \right], \tag{1.51}$$

where λ_i $(i = 1, 2, .., N)$ are the roots of the pseudo-polynomial $P(s^\alpha)$ or the system poles which are assumed to be simple without loss of generality. The analytical solution of the system (1.51) can be expressed as

$$y(t) = L^{-1} \left\{ K_0 \left[\sum_{i=1}^{N} \frac{A_i}{s^\alpha + \lambda_i} \right] \right\} = K_0 \sum_{i=1}^{N} A_i t^\alpha E_{\alpha,\alpha}(-\lambda_i t^\alpha), \tag{1.52}$$

where $E_{\mu,\nu}(z)$ is the Mittag-Leffler function defined as (1.56).

A fractional order plant to be controlled can be described by a typical n-term linear homogeneous fractional order differential equation (FODE) in time domain

$$a_n D_t^{\alpha_n} y(t) + \cdots + a_1 D_t^{\alpha_1} y(t) + a_0 D_t^{\alpha_0} y(t) = 0 \tag{1.53}$$

where a_k $(k = 0, 1, \cdots, n)$ are constant coefficients of the FODE; α_k $(k = 0, 1, 2, \cdots, n)$ are real numbers. Without loss of generality, assume that $\alpha_n > \alpha_{n-1} > \ldots > \alpha_0 \geq 0$.

The analytical solution of the FODE (1.53) is given by the general formula in the form [Podlubny (1999a)]:

$$y(t) = \frac{1}{a_n} \sum_{m=0}^{\infty} \frac{(-1)^m}{m!} \sum_{\substack{k_0+k_1+\ldots+k_{n-2}=m \\ k_0 \geq 0; \ldots, k_{n-2} \geq 0}} (m; k_0, k_1, \ldots, k_{n-2})$$

$$\times \prod_{i=0}^{n-2} \left(\frac{a_i}{a_n}\right)^{k_i} \mathcal{E}_m \left(t, -\frac{a_{n-1}}{a_n}; \alpha_n - \alpha_{n-1}, \alpha_n\right.$$

$$\left. + \sum_{j=0}^{n-2} (\alpha_{n-1} - \alpha_j)k_j + 1\right), \tag{1.54}$$

where $(m; k_0, k_1, \ldots, k_{n-2})$ are the multinomial coefficients and $\mathcal{E}_k(t, y; \mu, \nu)$ is the function of Mittag-Leffler type introduced by Podlubny [Podlubny (1999a)]. The function is defined by

$$\mathcal{E}_k(t, y; \mu, \nu) = t^{\mu k + \nu - 1} E_{\mu,\nu}^{(k)}(y t^{\mu}), \quad (k = 0, 1, 2, \ldots), \tag{1.55}$$

where $E_{\mu,\nu}(z)$ is the Mittag-Leffler function of two parameters [Gorenflo (2004)]:

$$E_{\mu,\nu}(z) = \sum_{i=0}^{\infty} \frac{z^i}{\Gamma(\mu i + \nu)}, \quad (\mu > 0, \quad \nu > 0), \tag{1.56}$$

where e.g. $E_{1,1}(z) = e^z$, and where its k-th derivative is given by

$$E_{\mu,\nu}^{(k)}(z) = \sum_{i=0}^{\infty} \frac{(i+k)! \, z^i}{i! \, \Gamma(\mu i + \mu k + \nu)}, \quad (k = 0, 1, 2, \ldots). \tag{1.57}$$

Consider a control function which acts on the FODE system (1.53) as follows:

$$a_n D_t^{\alpha_n} y(t) + \cdots + a_1 D_t^{\alpha_1} y(t) + a_0 D_t^{\alpha_0} y(t) = u(t). \tag{1.58}$$

By Laplace transform, we can get a fractional transfer function:

$$G(s) = \frac{Y(s)}{U(s)} = \frac{1}{a_n s^{\alpha_n} + \cdots + a_1 s^{\alpha_1} + a_0 s^{\alpha_0}}. \tag{1.59}$$

The fractional order linear time invariant (LTI) system can also be represented by the following state-space model

$$_0D_t^{\mathbf{q}} x(t) = \mathbf{A}x(t) + \mathbf{B}u(t)$$

$$y(t) = \mathbf{C}x(t), \tag{1.60}$$

where $x \in \mathbb{R}^n$, $u \in \mathbb{R}^r$ and $y \in \mathbb{R}^p$ are the state, input and output vectors of the system and $\mathbf{A} \in \mathbb{R}^{n \times n}$, $\mathbf{B} \in \mathbb{R}^{n \times r}$, $\mathbf{C} \in \mathbb{R}^{p \times n}$, and $\mathbf{q} = [q_1, q_2, \ldots, q_n]^T$

are the fractional orders. If $q_1 = q_2 = \ldots q_n \equiv \alpha$, system (1.60) is called a commensurate order system, otherwise it is an incommensurate order system.

State transition matrix is

$$\mathbf{x}(t) = \left[\mathbf{I} + \frac{\mathbf{A}\mathbf{x}(0)}{\Gamma(1+\alpha)}t^\alpha + \frac{\mathbf{A}^2\mathbf{x}(0)}{\Gamma(1+2\alpha)}t^{2\alpha} + \ldots + \frac{\mathbf{A}^k\mathbf{x}(0)}{\Gamma(1+k\alpha)}t^{k\alpha} + \ldots \right]$$

$$= \left(\sum_{k=0}^{\infty} \frac{\mathbf{A}^k t^{k\alpha}}{\Gamma(1+k\alpha)} \right) \mathbf{x}(0) = \mathbf{\Phi}(t)\mathbf{x}(0). \qquad (1.61)$$

A fractional-order system described by n-term fractional differential equation (1.58) can be rewritten into the state-space representation in the form [Dorčák (2002)], [Yang (2006)]:

$$\begin{bmatrix} {}_0D^{q_1}x_1(t) \\ {}_0D^{q_2}x_2(t) \\ \cdot \\ \cdot \\ {}_0D^{q_n}x_n(t) \end{bmatrix} = \begin{bmatrix} 0 & 1 & . & . & 0 \\ 0 & 0 & 1 & . & 0 \\ \cdot & & \cdot & \cdot & \cdot \\ \cdot & & & \cdot & \cdot \\ -a_0/a_n & -a_1/a_n & . & . & a_{n-1}/a_n \end{bmatrix} \begin{bmatrix} x_1(t) \\ x_2(t) \\ \cdot \\ \cdot \\ x_n(t) \end{bmatrix} + \begin{bmatrix} 0 \\ 0 \\ \cdot \\ \cdot \\ 1/a_n \end{bmatrix} u(t)$$

$$y(t) = \begin{bmatrix} 1 & 0 & \ldots & 0 & 0 \end{bmatrix} \begin{bmatrix} x_1(t) \\ x_2(t) \\ \cdot \\ \cdot \\ x_n(t) \end{bmatrix}, \qquad (1.62)$$

where $\alpha_0 = 0$, $q_1 = \alpha_1$, $q_2 = \alpha_{n-1} - \alpha_{n-2}$, $\ldots q_n = \alpha_n - \alpha_{n-1}$, and with initial conditions:

$$x_1(0) = x_0^{(1)} = y_0, \quad x_2(0) = x_0^{(2)} = 0, \ldots$$

$$x_i(0) = x_0^{(i)} = \begin{cases} y_0^{(k)}, & \text{if } i = 2k + 1, \\ 0, & \text{if } i = 2k, \end{cases} \quad i \le n. \qquad (1.63)$$

The n-term FODE (1.58) is equivalent to the system of equations (1.62) with the initial conditions (1.63) if Caputo derivative is considered.

The controllability, just like the conventional observability and controllability concept, is defined as follows [Matignon (1996a)]: System (1.60) is *controllable* on $[t_0, t_{final}]$ if controllability matrix

$$C_a = [B|AB|A^2B|\ldots|A^{n-1}B]$$

has rank n.

The observability is defined as follows [Matignon (1996a)]: System (1.60) is *observable* on $[t_0, t_{final}]$ if observanility matrix

$$O_a = \begin{bmatrix} C \\ CA \\ CA^2 \\ \ldots \\ CA^{n-1} \end{bmatrix}$$

has rank n.

1.8   Fractional Nonlinear Systems

Generally, we consider the following incommensurate fractional order nonlinear system in the form:

$$\begin{aligned} {}_0D_t^{q_i}x_i(t) &= f_i(x_1(t), x_2(t), \ldots, x_n(t), t) \\ x_i(0) &= c_i, \quad i = 1, 2, \ldots, n, \end{aligned} \qquad (1.64)$$

where c_i are initial conditions, or in its vector representation:

$$D^\mathbf{q}\mathbf{x} = \mathbf{f}(\mathbf{x}), \qquad (1.65)$$

where $\mathbf{q} = [q_1, q_2, \ldots, q_n]^T$ for $0 < q_i < 2$, $(i = 1, 2, \ldots, n)$ and $\mathbf{x} \in \mathbb{R}^n$.

The equilibrium points of the system (3.2) are calculated via solving the following equation

$$\mathbf{f}(\mathbf{x}) = 0 \qquad (1.66)$$

and we suppose that $x^* = (x_1^*, x_2^*, \ldots, x_n^*)$ is an equilibrium point of the system (3.2).

1.9   Stability of Fractional LTI Systems

Just as what has been considered in the previous subsection, even in the fractional case, the stability is different from that in the integer one. An interesting point is that a stable fractional system may have roots in right half of the complex w-plane. Since the principal sheet of the Riemann surface is defined $-\pi < \arg(s) < \pi$, by using the mapping $w = s^q$, the corresponding w domain is defined by $-q\pi < \arg(w) < q\pi$, and the w plane region corresponding to the right half plane of this sheet is defined by $-q\pi/2 < \arg(w) < q\pi/2$.

Consider the fractional order pseudo-polynomial

$$Q(s) = a_1 s^{q_1} + a_2 s^{q_2} + \ldots + a_n s^{q_n} = a_1 s^{c_1/d_1} + a_2 s^{c_2/d_2} + \ldots + a_n s^{c_n/d_n},$$

where q_i are rational number expressed as c_i/d_i and a_i are the real numbers for $i = 1, 2, \ldots, n$. If for some i, $c_i = 0$ then $d_i = 1$. Let v be the least common multiple (LCM) of $d_1, d_2, \ldots d_n$ denote as $v = \text{LCM}\{d_1, d_2, \ldots d_n\}$, then [Ghartemani (2008)]

$$\begin{aligned} Q(s) &= a_1 s^{\frac{v_1}{v}} + a_2 s^{\frac{v_2}{v}} + \ldots + a_n s^{\frac{v_n}{v}} \qquad (1.67) \\ &= a_1 (s^{\frac{1}{v}})^{v_1} + a_2 (s^{\frac{1}{v}})^{v_2} + \ldots + a_n (s^{\frac{1}{v}})^{v_n}. \end{aligned}$$

The fractional degree (FDEG) of the polynomial $Q(s)$ is defined as [Ghartemani (2008)]

$$\text{FDEG}\{Q(s)\} = \max\{v_1, v_2, \ldots, v_n\}.$$

The domain of definition for (1.67) is the Riemann surface with v Riemann sheets where origin is a branch point of order $v - 1$ and the branch cut is assumed at R^-. The number of roots for the fractional algebraic equation (1.67) is given by the following proposition [Bayat (2009)]:

Proposition 1.1. Let $Q(s)$ be a fractional order polynomial with $\text{FDEG}\{Q(s)\} = n$. Then the equation Q(s)=0 has exactly n roots on the Riemann surface.

Definition 1.1. The fractional order polynomial

$$Q(s) = a_1 s^{\frac{n}{v}} + a_2 s^{\frac{n-1}{v}} + \ldots + a_n s^{\frac{1}{v}} + a_{n+1}$$

is *minimal* if $\text{FDEG}\{Q(s)\} = n$. We will assume that all fractional order polynomials are minimal.

This ensures that there is no redundancy in the number of the Riemann sheets [Ghartemani (2008)].

On the other hand, it has been shown, by several authors and by using several methods, that for the case of FOLTI system of commensurate order, a geometrical method of complex analysis based on the argument principle of the roots of the characteristic equation (a polynomial in this particular case) can be used for the stability check in the BIBO sense (see e.g. [Matignon (1998)], [Petráš (1999a)]). The stability condition can then be stated as follows [Matignon (1996b, 1998)], [Vinagre (2007)]:

Theorem 1.1. *[Matignon (1996b)] A commensurate order system described by a rational transfer function (1.50) is stable if only if*

$$|\arg(\lambda_i)| > \alpha \frac{\pi}{2}, \text{ for all } i$$

with λ_i the i-th root of $P(s^\alpha)$.

In the case of the FOLTI system with commensurate order where the system poles are in general complex conjugate, the stability condition can also be expressed as follows [Matignon (1996b, 1998)]:

Theorem 1.2. *[Matignon (1998)] A commensurate order system described by a rational transfer function*

$$G(w) = \frac{Q(w)}{P(w)},$$

where $w = s^q$, $q \in R^+$, $(0 < q < 2)$, is stable if only if

$$|\arg(w_i)| > q\frac{\pi}{2},$$

with $\forall w_i \in C$ the i-th root of $P(w) = 0$.

When $w = 0$ is a single root (singularity at the origin) of P, the system cannot be stable. For $q = 1$, this is the classical theorem of pole location in the complex plane: it has no pole in the closed right half plane of the first Riemann sheet. The stability region suggested by this theorem tends to the whole s-plane when q tends to 0, corresponds to the Routh-Hurwitz stability when $q = 1$, and tends to the negative real axis when q tends to 2.

Theorem 1.3. *It has been shown that the commensurate system (1.60) is stable if the following condition is satisfied (also if the triplet A, B, C is minimal) [Aoun (2004); Matignon (1998); Tavazoei (2007a,b, 2008b)]:*

$$|arg(eig(\boldsymbol{A}))| > q\frac{\pi}{2}, \tag{1.68}$$

where $0 < q < 2$ and $eig(\boldsymbol{A})$ represents the eigenvalues of matrix $\boldsymbol{A}$.

Proposition 1.2. We can assume, that some incommensurate order systems described by the FODE (1.58) or (1.60), can be decomposed to the following modal form of the fractional transfer function (so called Laguerre functions [Aoun (2007)]):

$$F(s) = \sum_{i=1}^{N} \sum_{k=1}^{n_k} \frac{A_{i,k}}{(s^{q_i} + \lambda_i)^k} \tag{1.69}$$

for some complex numbers $A_{i,k}$, λ_i, and positive integer n_k.

A system (1.69) is BIBO stable if and only if q_i and the argument of λ_i denoted by $\arg(\lambda_i)$ in (1.69) satisfy the inequalities

$$0 < q_i < 2 \quad \text{and} \quad |\arg(\lambda_i)| < \pi\left(1 - \frac{q_i}{2}\right) \quad \text{for all } i. \tag{1.70}$$

Henceforth, we will restrict the parameters q_i to the interval $q_i \in (0,2)$. In the case $q_i = 1$ for all i we obtain a classical stability condition for the integer order system (no pole is in the right half plane). The inequalities (1.70) were obtained by applying the stability results given in [Akcay (2008)], [Matignon (1998)].

Theorem 1.4. *[Deng (2007c)] Consider the following autonomous system for internal stability definition:*

$$_0D_t^q x(t) = \boldsymbol{A}x(t), \quad x(0) = x_0, \tag{1.71}$$

with $\boldsymbol{q} = [q_1, q_2, \ldots, q_n]^T$ and its n-dimensional representation:

$$_0D_t^{q_1} x_1(t) = a_{11}x_1(t) + a_{12}x_2(t) + \cdots + a_{1n}x_n(t)$$
$$_0D_t^{q_2} x_2(t) = a_{21}x_1(t) + a_{22}x_2(t) + \cdots + a_{2n}x_n(t)$$
$$\ldots$$
$$_0D_t^{q_n} x_n(t) = a_{n1}x_1(t) + a_{n2}x_2(t) + \cdots + a_{nn}x_n(t) \tag{1.72}$$

where all q_i's are rational numbers between 0 and 2. Assume m be the LCM of the denominators u_i's of q_i's , where $q_i = v_i/u_i$, $v_i, u_i \in Z^+$ for $i = 1, 2, \ldots, n$ and we set $\gamma = 1/m$. Define:

$$\det \begin{pmatrix} \lambda^{mq_1} - a_{11} & -a_{12} & \cdots & -a_{1n} \\ -a_{21} & \lambda^{mq_2} - a_{22} & \cdots & -a_{2n} \\ \cdots & & & \\ -a_{n1} & -a_{n2} & \cdots & \lambda^{mq_n} - a_{nn} \end{pmatrix} = 0. \tag{1.73}$$

The characteristic equation (1.73) can be transformed to an integer order polynomial equation if all q_i's are rational numbers. Then the zero solution of system (1.72) is globally asymptotically stable if all roots λ_i's of the characteristic (polynomial) equation (1.73) satisfy

$$|arg(\lambda_i)| > \gamma \frac{\pi}{2} \text{ for all } i.$$

Denote λ by s^γ in equation (1.73), we get the characteristic equation in the form $\det(s^\gamma I - A) = 0$.

Corollary 1.1. Suppose $q_1 = q_2 = \ldots, q_n \equiv q$, $q \in (0,2)$, all eigenvalues λ of matrix A in (1.62) satisfy $|arg(\lambda)| > q\pi/2$, the characteristic equation becomes $\det(s^q I - A) = 0$ and all characteristic roots of the system (1.60) have negative real parts [Deng (2007c)]. This result is Theorem 1 of paper [Matignon (1996b)].

Remark 1.1. Generally, when we assume $s = |r|e^{i\phi}$, where $|r|$ is modulus and ϕ is argument of complex number in s-plane, respectively, transformation $w = s^{\frac{1}{m}}$ to complex w-plane can be viewed as $s = |r|^{\frac{1}{m}}e^{\frac{i\phi}{m}}$ and thus $|\arg(s)| = m.|\arg(w)|$ and $|s| = |w|^m$. Proof of this statement is obvious.

Stability analysis criteria for a general FOLTI system can be summarized as follow:

The characteristic equation of a general LTI fractional order system of the form:

$$a_n s^{\alpha_n} + \ldots + a_1 s^{\alpha_1} + a_0 s^{\alpha_0} \equiv \sum_{i=0}^{n} a_i s^{\alpha_i} = 0 \qquad (1.74)$$

may be rewritten as

$$\sum_{i=0}^{n} a_i s^{\frac{u_i}{v_i}} = 0$$

and transformed into w-plane

$$\sum_{i=0}^{n} a_i w^i = 0, \qquad (1.75)$$

with $w = s^{\frac{k}{m}}$, where m is the LCM of v_i. The procedure of stability analysis is (see e.g. [Radwan (2009)]):

(1) For given a_i calculate the roots of equation (1.75) and find the absolute phase of all roots $|\phi_w|$.

(2) Roots in the primary sheet of the w-plane which have corresponding roots in the s-plane can be obtained by finding all roots which lie in the region $|\phi_w| < \frac{\pi}{m}$ then applying the inverse transformation $s = w^m$ (see Remark 1.1.1). The region where $|\phi_w| > \frac{\pi}{m}$ is not physical. To test the roots in the desired region the matrix approach (1.76) can be used.

$$A_1 = \begin{bmatrix} A\cos\delta & -A\sin\delta \\ A\sin\delta & A\cos\delta \end{bmatrix} \equiv A \otimes \begin{bmatrix} \cos\delta & -\sin\delta \\ \sin\delta & \cos\delta \end{bmatrix} \qquad (1.76)$$

In fact a simple test can be used [Anderson (1974)]. Roots of polynomial $P(s) = \det(sI - A)$ lie inside in region $-\pi/2 - \delta < \arg(s) < \pi/2 + \delta$ if eigenvalues of the matrix have negative real part, where $\otimes$ denotes Kronecker product. This property has been used to stability analysis of ordinary fractional order LTI systems and also for interval fractional order LTI systems [Tavazoei (2009)].

(3) The condition for stability is $\frac{\pi}{2m} < |\phi_w| < \frac{\pi}{m}$. The condition for oscillation is $|\phi_w| = \frac{\pi}{2m}$ otherwise the system is unstable. If there is no root in the physical s-plane, the system will always be stable. [Radwan (2009)].

Example 1.1. : Let us consider the linear fractional order LTI system described by the transfer function [Dorčák (1994)], [Podlubny (1999a)]:

$$G(s) = \frac{Y(s)}{U(s)} = \frac{1}{0.8s^{2.2} + 0.5s^{0.9} + 1},\qquad(1.77)$$

and corresponding FODE has the following form:

$$0.8\,{}_0D_t^{2.2}y(t) + 0.5\,{}_0D_t^{0.9}y(t) + y(t) = u(t)\qquad(1.78)$$

with zero initial conditions.

The system (1.78) can be rewritten to its state space representation $(x_1(t) \equiv y(t))$:

$$\begin{bmatrix} {}_0D^{\frac{9}{10}}x_1(t) \\ {}_0D^{\frac{13}{10}}x_2(t) \end{bmatrix} = \begin{bmatrix} 0 & 1 \\ -1/0.8 & -0.5/0.8 \end{bmatrix}\begin{bmatrix} x_1(t) \\ x_2(t) \end{bmatrix} + \begin{bmatrix} 0 \\ 1/0.8 \end{bmatrix}u(t)$$

$$y(t) = \begin{bmatrix} 1 & 0 \end{bmatrix}\begin{bmatrix} x_1(t) \\ x_2(t) \end{bmatrix}\qquad(1.79)$$

The eigenvalues of the matrix $\mathbf{A}$ are $\lambda_{1,2} = -0.3125 \pm 1.0735j$ and then $|\arg(\lambda_{1,2})| = 1.8541$.

The analytical solution of the FODE (1.78) for $u(t) = 0$ obtained from general solution (1.54) has form:

$$y(t) = \frac{1}{0.8}\sum_{k=0}^{\infty}\frac{(-1)^k}{k!}\left(\frac{1}{0.8}\right)^k \mathcal{E}_k\left(t, -\frac{0.5}{0.8}; 2.2 - 0.9, 2.2 + 0.9k\right).\qquad(1.80)$$

In Fig. 1.8 the analytical solution of the FODE (1.78) is depicted where $u(t) = 0$. As we can see in the figure, the solution is stable because $\lim_{t\to\infty}y(t) = 0$. Let us investigate stability according to the previously described method. The corresponding characteristic equation of system is:

$$P(s) : 0.8s^{2.2} + 0.5s^{0.9} + 1 = 0 \;\Rightarrow\; 0.8s^{\frac{22}{10}} + 0.5s^{\frac{9}{10}} + 1 = 0,\qquad(1.81)$$

when $m = 10$, $w = s^{\frac{1}{10}}$ then the roots w_i's and their appropriate arguments of polynomial

$$P(w) : 0.8w^{22} + 0.5w^9 + 1 = 0\qquad(1.82)$$

are:

$w_{1,2} = -0.9970 \pm 0.1182j, |\arg(w_{1,2})| = 3.023;$
$w_{3,4} = -0.9297 \pm 0.4414j, |\arg(w_{3,4})| = 2.698;$
$w_{5,6} = -0.7465 \pm 0.6420j, |\arg(w_{5,6})| = 2.431;$
$w_{7,8} = -0.5661 \pm 0.8633j, |\arg(w_{7,8})| = 2.151;$
$w_{9,10} = -0.259 \pm 0.9625j, |\arg(w_{9,10})| = 1.834;$
$w_{11,12} = -0.0254 \pm 1.0111j, |\arg(w_{11,12})| = 1.595;$
$w_{13,14} = 0.3080 \pm 0.9772j, |\arg(w_{11,12})| = 1.265;$
$w_{15,16} = 0.5243 \pm 0.8359j, |\arg(w_{15,16})| = 1.010;$
$w_{17,18} = 0.7793 \pm 0.6795j, |\arg(w_{17,18})| = 0.717;$
$w_{19,20} = 0.9084 \pm 0.3960j, |\arg(w_{19,20})| = 0.411;$
$w_{21,22} = 1.0045 \pm 0.1684j, |\arg(w_{21,22})| = 0.1661;$

Physical significance roots are in the first Riemann sheet, which is expressed by relation $-\pi/m < \phi < \pi/m$, where $\phi = \arg(w)$. In this case they are

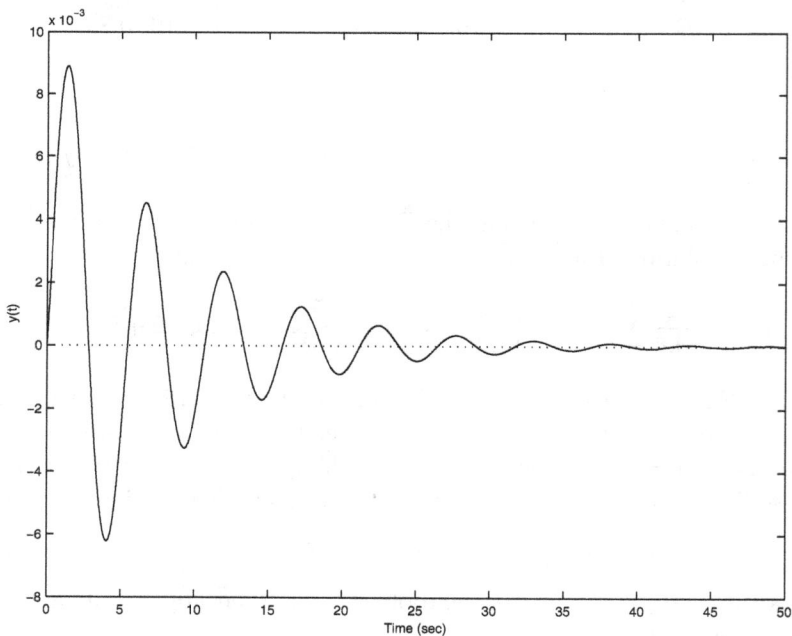

Fig. 1.8 Analytical solution of the FODE (1.78) where $u(t) = 0$ for $50\,s$ with zero initial conditions.

complex conjugate roots $w_{21,22} = 1.0045 \pm 0.1684j$ ($|\arg(w_{21,22})| = 0.1661$), which satisfy conditions $|\arg(w_{21,22})| > \pi/2m = \pi/20$. It means that the system (1.78) is stable (see Fig. 1.9). Other roots of the polynomial equation (1.82) lie in region $|\phi| > \frac{\pi}{m}$ which is not physical (outside of the closed angular sector limited by the thick line in Fig. 1.9(b)).

In Fig. 1.9(a) the Riemann surface of the function $w = s^{\frac{1}{10}}$ is depicted with the 10-Riemann sheets and in Fig. 1.9(b) the roots in complex w-plane are depicted with the angular sector corresponding to the stability region (dashed line) and the first Riemann sheet (thick line).

The interesting notion of Remark 1.1 should be mentioned here. The characteristic equation (1.81) has the following poles:

$$s_{1,2} = -0.10841 \pm 1.19699j,$$

in the first Riemann sheet in s-plane, which can be obtained e.g. via the Matlab routine as for instance:

```
>>s=solve('0.8*s^2.2+0.5*s^0.9+1=0','s')
```

When we compare $|\arg(w_{21,22})| = 0.1661$ and $|\arg(s_{1,2})| = 1.661$, we can see that $|\arg(s_{1,2})| = m|\arg(w_{21,22})|$, where $m = 10$ in transformation $w = s^{\frac{1}{m}}$. The first Riemann sheet is transformed from s-plane to w-plane as follows: $-\pi/10 < \arg(w) < \pi/10$ and in order to $-\pi < 10\arg(w) < \pi$. Therefore from this consideration we then obtain $|\arg(s)| = 10\,|\arg(w)|$.

Example 1.2. : Let us examine an interesting example of application, the so called Bessel function of the first kind, whose transfer function is [Matignon (1998)]:

$$H(s) = \frac{1}{\sqrt{s^2+1}} \quad \forall s,\ \Re(s) > 0. \tag{1.83}$$

We have two branch points $s_1 = i$, and $s_2 = -i$ and two cuts. One along the half line $(-\infty+i, i)$ and another one along the half line $(-\infty-i, -i)$. In this doubly cut complex plane, we have the identity $\sqrt{s^2+1} = \sqrt{s-i}\sqrt{s+i}$. The well known asymptotic expansion of equation (1.83) is:

$$h(t) \approx \sqrt{\frac{2}{\pi t}}\cos(t - \frac{\pi}{4}) = \sqrt{\frac{2}{\pi}} t^{-\frac{1}{2}} E_{2,1}\left(-(t - \frac{\pi}{4})^2\right).$$

According to the branch points and above asymptotic expansion we can state, that the system described by the Bessel function (1.83) is on the boundary of stability and has an oscillation behaviour.

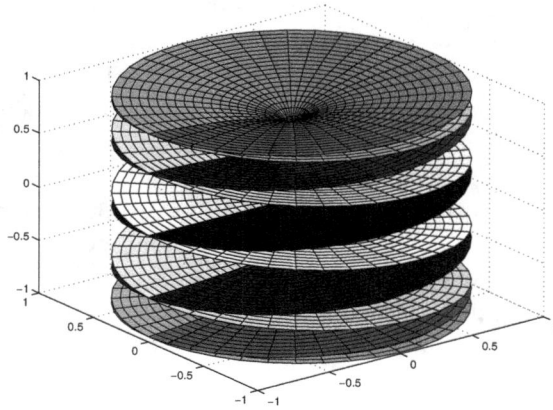

(a) 10-sheets Riemann surface

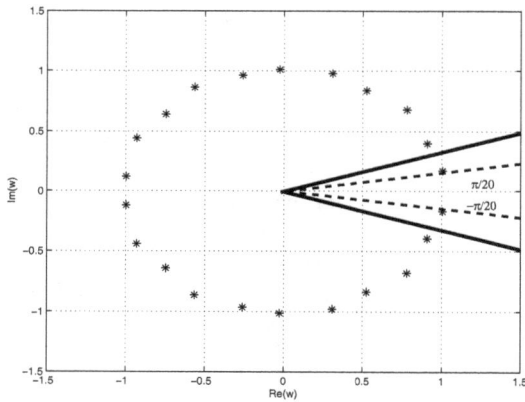

(b) Poles in complex w-plane

Fig. 1.9   Riemann surface of function $w = s^{\frac{1}{10}}$ and roots of equation (1.82) in complex w-plane.

Example 1.3. : Consider the closed loop system with the controlled system (electrical heater)

$$G(s) = \frac{1}{39.96s^{1.25} + 0.598} \qquad (1.84)$$

and PD controller

$$C(s) = 64.47 + 12.46s \tag{1.85}$$

The resulting closed loop transfer function $G_c(s)$ becomes [Petráš (2005b)]:

$$G_c(s) = \frac{Y(s)}{W(s)} = \frac{12.46s + 64.47}{39.69s^{1.25} + 12.46s + 65.068} \tag{1.86}$$

The analytical solution (impulse response) of the fractional order control system (1.86) is:

$$y(t) = \frac{12.46}{39.69} \sum_{k=0}^{\infty} \frac{(-1)^k}{k!} \left(\frac{12.46}{39.69}\right)^k \times \mathcal{E}_k\left(t, -\frac{65.068}{39.69}; 1.25, 0.25 - k\right)$$

$$+ \frac{64.47}{39.69} \sum_{k=0}^{\infty} \frac{(-1)^k}{k!} \left(\frac{65.068}{39.69}\right)^k \times \mathcal{E}_k\left(t, -\frac{12.46}{39.69}; 1.25 - 1, 1.25 + k\right)$$

$$\tag{1.87}$$

with zero initial conditions.

The characteristic equation of this system is

$$39.69s^{1.25} + 12.46s + 65.068 = 0 \quad \Rightarrow \quad 39.69s^{\frac{5}{4}} + 12.46s^{\frac{4}{4}} + 65.068 = 0 \tag{1.88}$$

Using the notation $w = s^{\frac{1}{m}}$, where LCM is $m = 4$, we obtain a polynomial of the complex variable w in form

$$39.69w^5 + 12.46w^4 + 65.068 = 0. \tag{1.89}$$

Solving the polynomial (1.89) we get the following roots and their arguments:

$$w_1 = -1.17474, |arg(w_1)| = \pi$$

$$w_{2,3} = -0.40540 \pm 1.0426j, |arg(w_{2,3})| = 1.9416$$

$$w_{4,5} = 0.83580 \pm 0.64536j, |arg(w_{4,5})| = 0.6575$$

This first Riemann sheet is defined as a sector in the w-plane within interval $-\pi/4 < arg(w) < \pi/4$. Complex conjugate roots $w_{4,5}$ lie in this interval and it satisfies the stability condition given as $|arg(w)| > \frac{\pi}{8}$, therefore the system is stable. The region where $|arg(w)| > \frac{\pi}{4}$ is not physical.

1.10   Stability of Fractional Nonlinear Systems

As it has been mentioned in [Matignon (1996b)], exponential stability cannot be used to characterize the asymptotic stability of fractional order systems. A new definition was introduced [Oustaloup (2008)].

Definition 1.2. Trajectory $x(t) = 0$ of the system (1.64) is t^{-q} asymptotically stable if there is a positive real q such that:

$$\forall ||x(t)|| \text{ with } t \le t_0, \, \exists N(x(t)), \text{ such that } \forall t \ge t_0, ||x(t)|| \le Nt^{-q}.$$

The fact that the components of $x(t)$ slowly decay towards 0 following t^{-q} leads to fractional systems sometimes being called long memory systems. Power law stability t^{-q} is a special case of the Mittag-Leffler stability [Li (2008)].

According to stability theorem defined in [Tavazoei (2009)], the equilibrium points are asymptotically stable for $q_1 = q_2 = \cdots = q_n \equiv q$ if all the eigenvalues λ_i, $(i = 1, 2, \ldots, n)$ of the Jacobian matrix $\mathbf{J} = \partial \mathbf{f}/\partial \mathbf{x}$, where $\mathbf{f} = [f_1, f_2, \ldots, f_n]^T$, evaluated at the equilibrium, satisfy the condition [Tavazoei (2007b)], [Tavazoei (2007a)]:

$$|\arg(\text{eig}(\mathbf{J}))| = |\arg(\lambda_i)| > q\frac{\pi}{2}, \quad i = 1, 2, \ldots, n. \tag{1.90}$$

Fig. 1.10 shows stable and unstable regions of the complex plane for such case.

Now, consider the incommensurate fractional order system $q_1 \ne q_2 \ne \cdots \ne q_n$ and suppose that m is the LCM of the denominators u_i's of q_i's, where $q_i = v_i/u_i$, $v_i, u_i \in Z^+$ for $i = 1, 2, \ldots, n$ and we set $\gamma = 1/m$. System (3.2) is asymptotically stable if:

$$|\arg(\lambda)| > \gamma\frac{\pi}{2}$$

for all roots λ of the following equation

$$\det(\text{diag}([\lambda^{mq_1} \lambda^{mq_2} \ldots \lambda^{mq_n}]) - \mathbf{J}) = 0. \tag{1.91}$$

A necessary stability condition for fractional order systems (3.2) to remain chaotic is keeping at least one eigenvalue λ in the unstable region [Tavazoei (2007b)]. The number of saddle points and eigenvalues for one-scroll, double-scroll and multi-scroll attractors was exactly described in work [Tavazoei (2008b)]. Assume that a 3D chaotic system has only three equilibria. Therefore, if system has double-scroll attractor, it has two saddle points surrounded by scrolls and one additional saddle point. Suppose

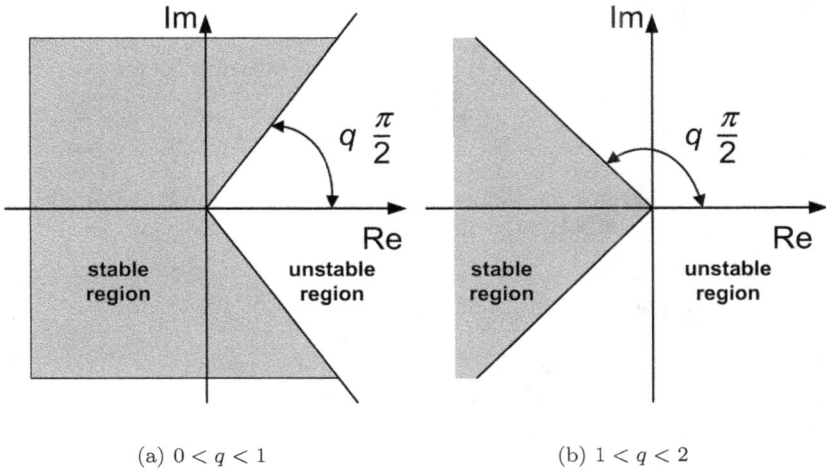

(a) $0 < q < 1$ (b) $1 < q < 2$

Fig. 1.10   Stability regions of the fractional order system.

that the unstable eigenvalues of scroll saddle points are: $\lambda_{1,2} = \alpha_{1,2} \pm j\beta_{1,2}$. The necessary condition to exhibit double-scroll attractor of system (3.2) is the eigenvalues $\lambda_{1,2}$ remaining in the unstable region [Tavazoei (2008b)]. The condition for commensurate derivatives order is

$$q > \frac{2}{\pi}\text{atan}\left(\frac{|\beta_i|}{\alpha_i}\right), \quad i = 1, 2. \tag{1.92}$$

This condition can be used to determine the minimum order for which a nonlinear system can generate chaos [Tavazoei (2007b)]. In other words, when the instability measure $\pi/2m - \min(|\text{arg}(\lambda)|)$ is negative, the system can not be chaotic.

Example 1.4. : Let us investigate the Chen system with a double scroll attractor. The fractional order form of such a system can be described as [Tavazoei (2008a)]

$$_0D_t^{0.8}x_1(t) = 35[x_2(t) - x_1(t)]$$
$$_0D_t^{1.0}x_2(t) = -7x_1(t) - x_1(t)x_3(t) + 28x_2(t)$$
$$_0D_t^{0.9}x_3(t) = x_1(t)x_2(t) - 3x_3(t) \tag{1.93}$$

The system has three equilibria at $(0,0,0)$, $(7.94, 7.94, 21)$, and $(-7.94, -7.94, 21)$. The Jacobian matrix of the system evaluated at (x_1^*, x_2^*, x_3^*) is:

$$\mathbf{J} = \begin{bmatrix} -35 & 35 & 0 \\ -7 - x_3^* & 28 & -x_1^* \\ x_2^* & x_1^* & -3 \end{bmatrix}. \tag{1.94}$$

The two last equilibrium points are saddle points and surrounded by a chaotic double scroll attractor. For these two points, equation (1.91) becomes as follows:

$$\lambda^{27} + 35\lambda^{19} + 3\lambda^{18} - 28\lambda^{17} + 105\lambda^{10} - 21\lambda^8 + 4410 = 0 \tag{1.95}$$

The characteristic equation (1.95) has unstable roots $\lambda_{1,2} = 1.2928 \pm 0.2032j$, $|\arg(\lambda_{1,2})| = 0.1560$ and therefore the system (1.93) satisfies the necessary condition to exhibit a double scroll attractor. The instability measure is 0.0012.

Numerical simulation of the system (1.93) for initial conditions $(-9, -5, 14)$ is depicted in Fig. 1.11.

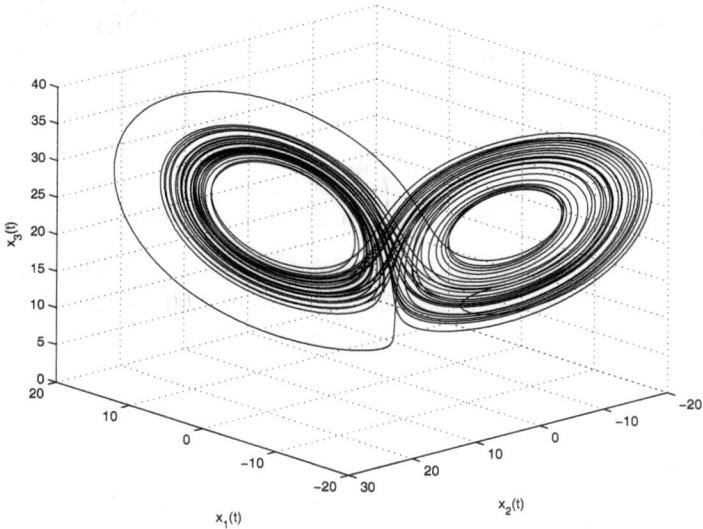

Fig. 1.11   Double scroll attractor of Chen's system (1.93) projected into 3D state space for 30 sec.

Chapter 2

Fractional Order PID Controller and their Stability Regions Definition

A new procedure that makes it possible to define the parameters of fractional order $PI^\lambda D^\mu$ controller, designed to stabilize a first-order plant with delay-time, is proposed in this chapter. The complete set of the stabilizing $PI^\lambda D^\mu$ parameters is determined using a version of the Hermite-Biehler theorem applicable to quasipolynomials. The widespread industrial use of PID controllers and the potentiality of their non integer order representation justifies a timely interest in $PI^\lambda D^\mu$ tuning techniques.

2.1 Introduction

Due to the absence of appropriate mathematical methods, fractional-order dynamical systems have only been studied marginally in the theory and practice of control systems. Some successful attempts have been undertaken but generally the study in the time domain has been almost avoided. However, in the last years a renewed interest has been devoted to fractional order systems in the area of automatic control.

It is possible to apply non integer order systems for control purposes as in [Oustaloup (1995)], [Podlubny (1999a)], [Podlubny (1999b)] and [Arena (2000)], and in robotic [Machado (2008)], while different practical controller implementations have been suggested in [Bohannan (2006)] and [Podlubny (2002a)].

The three CRONE control generations, CRONE being the French acronym of "Commande Robuste d'Ordre Non Entier" which means Robust Control of non integer order, represent the first framework for non integer order systems application in the automatic control area [Oustaloup (1995)], [Oustaloup (1993a)], [Oustaloup (1993b)] and [Oustaloup (1993c)].

Much interest is equally devoted to $PI^\lambda D^\mu$ parameters tuning, see for example [Vinagre (2006a)], [Valerio (2006)], [Vinagre (2006b)] and [Caponetto (2004)].

$PI^\lambda D^\mu$ has been introduced in [Podlubny (1999a)] and in the same paper a better response of this type of controller was demonstrated, in comparison with the classical PID, when used for the control of fractional order systems. A frequency domain approach by using $PI^\lambda D^\mu$ controllers is also studied in [Vinagre (2006b)].

The authors in [Caponetto (2006)] and [Caponetto (2008a)] proposed an analog implementation of the non integer order integrator based on Field Programmable Analog Arrays, (FPAAs), able to implement $PI^\lambda D^\mu$ controller.

All over the world a lot of control systems are operated by industrial PID controllers. Thanks to the widespread industrial use of PID controllers, even a small improvement in PID features, achieved by using $PI^\lambda D^\mu$, could have a relevant impact.

During the last decades, numerous methods have been developed for the setting of the parameters of P, PI, and PID controllers. Some of these methods are based on characterizing the dynamic response of the plant to be controlled by using a first-order model with time delay. It is interesting to note that even though most of these tuning techniques provide satisfactory results, the set of all stabilizing PID controllers for these first-order models with time delay remains unknown.

In an earlier work, [Bhattacharyya (2000)], a generalization of the Hermite-Biehler theorem was derived and was then used to compute the set of all stabilizing PID controllers for a given linear, time invariant plant, described by a rational transfer function. The approach developed in [Bhattacharyya (2000)] constitutes the first attempt to find a characterization of all stabilizing PID controllers for a given plant. However, the synthesis results presented in that reference cannot be applied directly to plants containing time delays since these were obtained for plants described by rational transfer functions.

Plants with time delays give rise to characteristic equations containing quasipolynomials.

In a successive work [Bhattacharyya (2002)] the problem of characterizing the set of classical PID controller parameters that stabilizes a given first-order plant with time delay was handled. In this paper a version of the Hermite-Biehler theorem applicable to quasipolynomials has been presented deriving results from the Pontryagin theory given in [Pontryagin (1995)] and [Karmarkar (1970)].

The aim of this paper is to extend the results given in [Bhattacharyya (2002)] to Fractional Order Systems (FOS) by providing a complete solution to the problem of characterizing the set of the gains of the non integer order $PI^{\lambda}D^{\mu}$ controller, fixing previously the fractional orders of the integrative (λ) and derivative (μ) actions, that make the closed-loop system stable.

2.2 Problem Characterization

Systems with step responses like the one shown in Fig. 2.1 are commonly modelled as first order processes with a time delay, and can be mathematically described by

$$G(s) = \frac{k}{1+Ts}e^{-Ls} \qquad (2.1)$$

where k represents the steady-state gain of the plant, L represents the time delay, and T represents the time constant of the plant.

The feedback control system shown in Fig. 2.2 is now considered where u is the command signal, y is the plant output, $G(s)$, given by (2.1), is the plant to be controlled, and C(s) is the control system.

In [Bhattacharyya (2002)] the problem of stabilizing a first order plant with time delay, $C(s)$ being a PID controller, was addressed. The range of admissible proportional gains is determined in a closed form for both open-loop stable and unstable plants. For each proportional gain in this

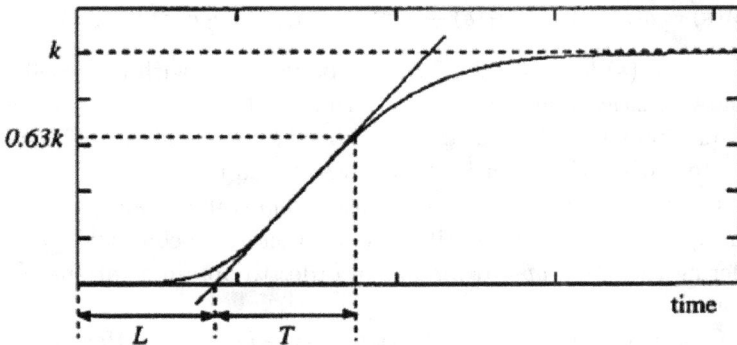

Fig. 2.1   Open-loop step response.

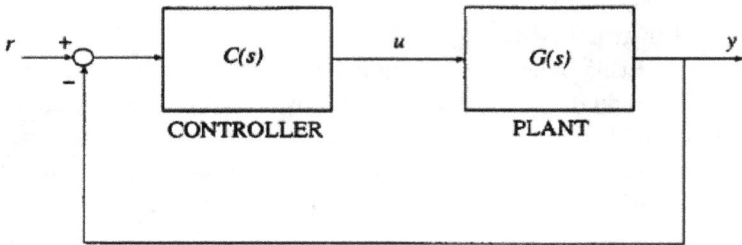

Fig. 2.2 Feedback control system.

range the stabilizing set in the space of the integral and derivative gains proves to be either a trapezoid, a triangle or a quadrilateral.

This chapter focuses on the case when the controller $C(s)$ is a fractional order $PI^\lambda D^\mu$ controller, i.e.:

$$C(s) = K_p + \frac{K_i}{s^\lambda} + K_d s^\mu \qquad (2.2)$$

The aim is to determine the set of controller parameters (K_p, K_i, K_d) for which the closed-loop system is stable [Caponetto (2008c)].

2.3 Theory for Analyzing Systems with Time Delays

Many problems in process control engineering involve time delays. These time delays lead to dynamic models with characteristic equations of the form

$$\delta(s) = d(s) + e^{-sT_1} n_1(s) + e^{-sT_2} n_2(s) + \ldots + e^{-sT_m} n_m(s) \qquad (2.3)$$

where $d(s)$, $n_i(s)$ for $i = 1, 2, \ldots, m$ are polynomials with real coefficients.

Characteristic equations of this form, in the variables s and e^s, are called quasipolynomials and are used to study the stability of the closed loop system when the plant is characterized by a time delay.

It is more effective to study the closed loop stability analyzing directly the quasipolynomial that describes the characteristic equation rather than considering the *Padé approximation* in order to obtain a rational transfer function.

In [Gantmacher (1959)] and [Bhattacharyya (1995)] the Hermite-Biehler theorem for Hurwitz polynomials was shown not to carry over to arbitrary functions of the complex variable s. In order to overcome this problem in

[Pontryagin (1995)] Pontryagin studied functions of the form $P(s, e^s)$ as that in (2.3).

Successively in [Bhattacharyya (1995)] and [Bellman (1963)], based on Pontryagin's results, an extension of the Hermite-Biehler theorem was developed to study the stability of a certain class of quasipolynomials.

In particular, this class is characterized by the following assumptions:

A1) $\deg[d(s)] = n$ and $\deg[n_i(s)] \leq n$ for $i = 1, 2, \ldots, m$
A2) $0 < T_1 < T_2 < \ldots < T_m$

The previous assumptions allow one to define a quasipolynomial with a non-zero principal term, i.e, the coefficient of the term containing the highest powers of s and e^s is different to zero.

In fact the quasipolynomials with no principal term are characterized by an infinite number of roots with positive real parts that make the closed loop system unstable. Quasipolynomials with principal term different from zero could be stable or unstable but with a finite number of positive roots [Pontryagin (1995)].

In the following, the study of the roots of quasipolynomials with the principal term different from zero will make it possible to determine the stability parameters region of the closed loop system, as that of Fig. 2.2, characterized by the non fractional controller given in equation (2.2).

From equation (2.3), since e^{sT_m} has not any finite zeros, the zeros of $\delta(s)$ are identical to those of $\delta^*(s)$ and the quasipolynomial assumes the following form:

$$\delta^*(s) = e^{sT_m}\delta(s) \tag{2.4}$$
$$= e^{sT_m}d(s) + e^{s(T_m-T_1)}n_1(s) + e^{s(T_m-T_2)}n_2(s) + \ldots + n_m(s)$$

The stability of the system having the characteristic equation (2.3) is therefore equivalent to the condition that all the zeros of $\delta^*(s)$ lie in the open left half-plane [Kharitonov (1994)].

The following theorem gives necessary and sufficient conditions for the stability of $\delta^*(s)$, see [Bellman (1963)].

2.3.1 Hermite-Biehler Theorem

Let $\delta^*(s)$ given by (2.4) in the form:

$$\delta^*(j\omega) = \delta_r^*(\omega) + j\delta_i^*(\omega) \tag{2.5}$$

where $\delta_r^*(\omega)$ and $\delta_i^*(\omega)$ represent respectively the real and imaginary parts of $\delta^*(j\omega)$. Under assumptions A1 and A2, $\delta^*(j\omega)$ is stable if and only if:

(1) $\delta_r^*(\omega)$ and $\delta_i^*(\omega)$ have only simple real roots and these are interlaced;

(2) $\delta'^*_i(\overline{\omega})\delta_r^*(\overline{\omega}) - \delta_i^*(\overline{\omega})\delta'^*_r(\overline{\omega}) > 0$, for some $\omega = \overline{\omega}$ in $(-\infty, +\infty)$,

where $\delta'^*_r(\omega)$ and $\delta'^*_i(\omega)$ denote the derivative with respect to ω of $\delta_r^*(\omega)$ and $\delta_i^*(\omega)$, respectively.

The second condition is the analytic formulation of the assumption that the vector $\delta^*(j\omega)$, as ω varies from $-\infty$ to $+\infty$, continuously rotates in the positive direction with positive velocity.

A crucial step in applying the above theorem is to ensure that $\delta_r^*(\omega)$ and $\delta_i^*(\omega)$ have only real roots. Such a property can be ensured by using the following result, also due to Pontryagin [Bellman (1963)].

2.3.2 *Pontryagin Theorem*

Let M and N denote the highest powers of s and e^s respectively in $\delta^*(s)$. Let η be an appropriate constant such that the coefficients of terms of highest degree in $\delta_r^*(\omega)$ and $\delta_i^*(\omega)$ do not vanish at $\omega = \eta$. Then in order to the equations $\delta_r^*(\omega) = 0$ and $\delta_i^*(\omega) = 0$ have only real roots, it is necessary and sufficient that in the intervals

$$-2l\pi + \eta \leq \omega \leq 2l\pi + \eta \qquad l = 1, 2, 3, \ldots \qquad (2.6)$$

$\delta_r^*(\omega)$ and $\delta_i^*(\omega)$ have exactly $4lN + M$ roots, starting with a sufficiently large l.

In the case of the characteristic equation with fractional order, i.e., M and N are non-integer numbers, $\delta_r^*(\omega)$ and $\delta_i^*(\omega)$ must have exactly $4l([N] + 1) + [M] + 1$ roots, where $[\ldots]$ means the integer part.

2.4 **Stability Regions with $PI^\lambda D^\mu$ Controller**

The closed-loop characteristic equation of the system in Fig. 2.2 with $C(s)$ given in (2.2) is:

$$\delta(s) = (kK_i + kK_p s^\lambda + kK_d s^{\lambda+\mu})e^{-Ls} + (1 + Ts)s^\lambda \qquad (2.7)$$

Theorems 2.3.1 and 2.3.2 are now applied to solve the stability problem and find the set of all stabilizing $PI^\lambda D^\mu$ controllers.

Let us start by rewriting the quasipolynomial $\delta(s)$ as

$$\delta^*(s) = kK_i + kK_p s^\lambda + kK_d s^{\lambda+\mu} + (1 + Ts)s^\lambda e^{Ls} = n(s) + d(s)e^{Ls} \qquad (2.8)$$

where $\deg[d(s)] = \lambda + 1$ and $\deg[n(s)] = \lambda + \mu$. In order to verify (A1) and (A2), the condition $\lambda + \mu \leq \lambda + 1$, that implies $\mu \leq 1$, and the condition $L > 0$, must be verified.

Given $\lambda = a/b$ and $\mu = c/b$, the previous condition implies $c \leq b$ and the equation (2.8) can be rewritten as:

$$\delta^*(s) = s^{a/b}\left[kK_d s^{c/b} + kK_p + (1 + Ts)e^{Ls}\right] + kK_i \qquad (2.9)$$

With the change of variable $z = Ls$, the quasipolynomial assumes the form:

$$\delta^*(z) = \left(\frac{z}{L}\right)^{a/b}\left[kK_d\left(\frac{z}{L}\right)^{c/b} + kK_p + (1 + \frac{T}{L}z)e^{z}\right] + kK_i \qquad (2.10)$$

Then for $z = j\omega$, $\delta^*(j\omega)$ becomes:

$$\delta^*(j\omega) = \left(\frac{j\omega}{L}\right)^{\frac{a+c}{b}} kK_d + \left(\frac{j\omega}{L}\right)^{\frac{a}{b}} kK_p + \left(\frac{j\omega}{L}\right)^{\frac{a}{b}}(1 + j\omega\frac{T}{L})e^{j\omega} + kK_i \qquad (2.11)$$

that is:

$$\delta^*(j\omega) = \left(\frac{j\omega}{L}\right)^{\frac{a+c}{b}} kK_d + \left(\frac{j\omega}{L}\right)^{\frac{a}{b}} kK_p + \left(\frac{j\omega}{L}\right)^{\frac{a}{b}}$$
$$* \left(\cos\omega - \frac{T}{L}\omega\sin\omega + j\sin\omega + j\frac{T}{L}\omega\cos\omega\right) + kK_i \qquad (2.12)$$

being $e^{j\omega} = \cos\omega + j\sin\omega$.

The real and imaginary part $\delta_r^*(\omega)$ and $\delta_i^*(\omega)$ are:

$$\delta_r^*(\omega) = kK_i + kK_d\left|\text{Re}\left\{(j)^{\frac{a+c}{b}}\right\}\right|\,|\omega|^{\frac{a+c}{b}}\left(\frac{1}{L}\right)^{\frac{a+c}{b}}$$
$$+ \left(kK_p + \cos\omega - \frac{T}{L}\omega\sin\omega\right)\left|\text{Re}\left\{(j)^{\frac{a}{b}}\right\}\right||\omega|^{\frac{a}{b}}\left(\frac{1}{L}\right)^{\frac{a}{b}}$$
$$- \left(\sin\omega + \frac{T}{L}\omega\cos\omega\right)\left|\text{Im}\left\{(j)^{\frac{a}{b}}\right\}\right||\omega|^{\frac{a}{b}}\left(\frac{1}{L}\right)^{\frac{a}{b}}\text{sign}(\omega) \qquad (2.13)$$

$$\delta_i^*(\omega) = kK_d\left|\text{Im}\left\{(j)^{\frac{a+c}{b}}\right\}\right||\omega|^{\frac{a+c}{b}}\left(\frac{1}{L}\right)^{\frac{a+c}{b}}\text{sign}(\omega)$$
$$+ \left(kK_p + \cos\omega - \frac{T}{L}\omega\sin\omega\right)\left|\text{Im}\left\{(j)^{\frac{a}{b}}\right\}\right||\omega|^{\frac{a}{b}}\left(\frac{1}{L}\right)^{\frac{a}{b}}$$
$$* \text{sign}(\omega) + \left(\sin\omega + \frac{T}{L}\omega\cos\omega\right)\left|\text{Re}\left\{(j)^{\frac{a}{b}}\right\}\right||\omega|^{\frac{a}{b}}\left(\frac{1}{L}\right)^{\frac{a}{b}} \qquad (2.14)$$

The factor $(j)^{\frac{a+c}{b}}$ as well as $(j)^{\frac{a}{b}}$ have b complex solutions, but the solution which must be considered in the characteristic equation is the common one with the smallest positive phase, as will be shown in the next section.

It is possible to note that, fixing the fractional integrative and derivative orders, i.e. a, b and c, the real part $\delta_r^*(\omega)$ depends on all the three parameters K_p, K_i and K_d, while the imaginary part $\delta_i^*(\omega)$ depends on two parameters K_p and K_d.

As previously stated, according to the Hermite-Biehler theorem it is possible to use the Pontryagin theorem to ensure that $\delta_r^*(\omega)$ and $\delta_i^*(\omega)$ have only real roots. Besides, if one part has only real roots and the interlacing property between the roots of the two parts is verified this implies that also the other part has only real roots [Bhattacharyya (2002)]. By taking into account these considerations it is sufficient to apply the Pontryagin theorem only to one part and it results more convenient to apply it on to the part that depends on only two parameters, that is, $\delta_i^*(\omega)$.

The successive step is to determine the conditions that make it possible to verify that the roots of $\delta_r^*(\omega)$ and $\delta_i^*(\omega)$ are interlaced. In order to verify this property the values of the real part $\delta_r^*(\omega)$ are calculated in the zeros of the imaginary part $\delta_i^*(\omega)$.

For $\omega \neq 0$ the real part $\delta_r^*(\omega)$ can be rewritten as

$$\delta_r^*(\omega) = k \left| \mathrm{Re}\left\{ (j)^{\frac{a}{b}} \right\} \right| |\omega|^{\frac{a}{b}} \left(\frac{1}{L} \right)^{\frac{a}{b}} * \left[K_p + m(\omega)K_d + n(\omega)K_i + b(\omega) \right]$$

$$(2.15)$$

where

$$m(\omega) = \frac{\left| \mathrm{Re}\left\{ (j)^{\frac{a+c}{b}} \right\} \right|}{\left| \mathrm{Re}\left\{ (j)^{\frac{a}{b}} \right\} \right|} |\omega|^{\frac{c}{b}} \left(\frac{1}{L} \right)^{\frac{c}{b}}, \qquad (2.16)$$

$$n(\omega) = \frac{1}{k} \left| \mathrm{Re}\left\{ (j)^{\frac{a}{b}} \right\} \right|^{-1} |\omega|^{-\frac{a}{b}} \left(\frac{1}{L} \right)^{-\frac{a}{b}}, \qquad (2.17)$$

and

$$b(\omega) = \frac{1}{k} \left[\left(\cos\omega - \frac{T}{L}\omega \sin\omega \right) - \frac{\left| \mathrm{Im}\left\{ (j)^{\frac{a}{b}} \right\} \right|}{\left| \mathrm{Re}\left\{ (j)^{\frac{a}{b}} \right\} \right|} \left(\sin\omega + \frac{T}{L}\omega \cos\omega \right) \mathrm{sign}(\omega) \right]$$

$$(2.18)$$

Since $\delta_i^*(\omega)$ is a odd function, it always has a root in $\omega = 0$. Thus for $\omega = \omega_0 = 0$

$$\delta_r^*(\omega_0) = kK_i \tag{2.19}$$

while for $\omega = \omega_j \neq 0$

$$\delta_r^*(\omega_j) = k\left|\mathrm{Re}\left\{(j)^{\frac{a}{b}}\right\}\right|\left|\omega_j\right|^{\frac{a}{b}}\left(\frac{1}{L}\right)^{\frac{a}{b}} * \left[K_p + m(\omega_j)K_d + n(\omega_j)K_i + b(\omega_j)\right] \tag{2.20}$$

Thus, in order to verify the interlace property between the roots of $\delta_r^*(\omega)$ and $\delta_i^*(\omega)$ one must impose

$$\delta_r^*(\omega_0) \gtrless 0 \Rightarrow K_i \gtrless 0 \tag{2.21}$$

and

$$(-1)^j \delta_r^*(\omega_j) \gtrless 0 \tag{2.22}$$
$$\Rightarrow (-1)^j m(\omega_j)K_d + (-1)^j n(\omega_j)K_i + (-1)^j b(\omega_j) \gtrless (-1)^{j+1} K_p$$

taking into account that the term $k\left|\mathrm{Re}\left\{(j)^{\frac{a}{b}}\right\}\right|\left|\omega\right|^{\frac{a}{b}}\left(\frac{1}{L}\right)^{\frac{a}{b}}$ is positive because $k > 0$ and $L > 0$.

These conditions related to the real part define a volume in the space (K_p, K_i, K_d), as will be shown in the next section.

The last condition of the Hermite-Biehler theorem, $\delta'^*_i(\overline{\omega})\delta_r^*(\overline{\omega}) - \delta_i^*(\overline{\omega})\delta'^*_r(\overline{\omega}) > 0$, for some $\omega = \overline{\omega}$ in $(-\infty, +\infty)$, defines another volume in the space.

By taking into account the intersection of the two volumes, the region on the space (K_p, K_i, K_d), where the closed loop system is stable, is defined.

2.5   Results

In order to calculate the set of parameters (K_p, K_i, K_d), which ensure the stability of the closed loop system, three procedures have been developed in a Matlab Environment.

The first one implements the Pontryagin theorem defining the range of the values of K_p and K_d which ensure real roots for $\delta_i^*(\omega)$ according to the value of η (see eq. (2.6)). This procedure allows one to make it possible to define the value of η in order to obtain the widest range of the parameters K_p and K_d, otherwise unknown.

The second procedure defines the set of the values of the parameters (K_p, K_i, K_d) that verifies the interlace property of the roots of $\delta_r^*(\omega)$ or $\delta_i^*(\omega)$.

The third procedure defines the volume of parameters (K_p, K_i, K_d) that fulfil the last condition of the Hermite-Biehler theorem.

In order to explain the three developed procedures firstly the case of a fractional PD controller is considered. Thus $K_i = 0$ and $a = 0$ and, for example, a derivative action of order equal to $\mu = 1/3$ are selected. Besides the plant parameters are fixed as $k = 1$, $T = 2$ and $L = 1.2$ for example. The aim is to obtain the complete range of values of the parameters K_p and K_d that make the closed loop system stable.

By taking $K_i = 0$ and $a = 0$ the characteristic equation (2.12) becomes:

$$\delta^*(j\omega) = \left(\frac{j\omega}{L}\right)^{\frac{c}{b}} kK_d + kK_p + \left(\cos\omega - \frac{T}{L}\omega\sin\omega + j\sin\omega + j\frac{T}{L}\omega\cos\omega\right)$$

$$(2.23)$$

and the real and imaginary parts become:

$$\delta_r^*(\omega) = kK_d\left|\mathrm{Re}\left\{(j)^{\frac{c}{b}}\right\}\right||\omega|^{\frac{c}{b}}\left(\frac{1}{L}\right)^{\frac{c}{b}} + \left(kK_p + \cos\omega - \frac{T}{L}\omega\sin\omega\right)$$

$$(2.24)$$

$$\delta_i^*(\omega) = kK_d\left|\mathrm{Im}\left\{(j)^{\frac{c}{b}}\right\}\right||\omega|^{\frac{c}{b}}\left(\frac{1}{L}\right)^{\frac{c}{b}}\mathrm{sign}(\omega) + \left(\sin\omega + \frac{T}{L}\omega\cos\omega\right)$$

$$(2.25)$$

The factor $(j)^{\frac{c}{b}}$ has b complex solutions, but the solution which must be considered in the characteristic equation is the common one with the smallest positive phase. In fact when selecting for example $c = 1$ and $b = 3$, $(j)^{\frac{1}{3}}$ has 3 complex solutions equal to: $-j$ and $\pm\frac{\sqrt{3}}{2} + j0.5$. But a derivative action $\mu = \frac{1}{3}$ is also obtained by selecting $c = 2$ and $b = 6$, or $c = 3$ and $b = 9$, or $c = 4$ and $b = 12$, and so on. In Tab. 2.1 the solutions of $(j)^{0.\overline{3}}$ are shown. The common solution with the smallest positive phase is $\frac{\sqrt{3}}{2} + j0.5$.

By replacing this value to the real and imaginary part of $(j)^{0.\overline{3}}$, $\delta_r^*(\omega)$ and $\delta_i^*(\omega)$ become:

$$\delta_r^*(\omega) = kK_d\frac{\sqrt{3}}{2}|\omega|^{\frac{1}{3}}\left(\frac{1}{L}\right)^{\frac{1}{3}} + kK_p + \cos\omega - \frac{T}{L}\omega\sin\omega \qquad (2.26)$$

$$\delta_i^*(\omega) = kK_d\frac{1}{2}|\omega|^{\frac{1}{3}}\left(\frac{1}{L}\right)^{\frac{1}{3}}\mathrm{sign}(\omega) + \sin\omega + \frac{T}{L}\omega\cos\omega \qquad (2.27)$$

Thus $\delta_i^*(\omega)$ only depends on the parameter K_d, while $\delta_r^*(\omega)$ depends on the two parameters K_p and K_d.

By applying the first algorithm to $\delta_i^*(\omega)$ the range of the values of K_d, which ensures real roots for $\delta_i^*(\omega)$ according to the value of η, is found and shown in Fig. 2.3. Thanks to this algorithm it is possible to choose the value of η that ensures the widest range of K_d, $(-3.7 < K_d < 8.1$ in this case).

By fixing, for example, $\eta = 1.1$ in Fig. 2.4 it is possible to note that out of the range $-3.7 < K_d < 8.1$, $\delta_i^*(\omega)$ has not 5 roots in the interval $-2\pi + \eta \le \omega \le 2\pi + \eta$, neither $4l([N] + 1) + [M] + 1$ roots in $-2l\pi + \eta \le \omega \le 2l\pi + \eta$ for $l = 1, 2, 3, \ldots$.

Table 2.1   Complex solutions of $(j)^{\frac{1}{3}}$

$PD^{1/3}$	$PD^{2/6}$	$PD^{3/9}$	$PD^{4/12}$
$-j$	$\pm j$	$-j$	$\pm j$
$\pm\frac{\sqrt{3}}{2} + j0.5$	$\pm\frac{\sqrt{3}}{2} + j0.5$	$\pm\frac{\sqrt{3}}{2} + j0.5$	± 1
	$\pm\frac{\sqrt{3}}{2} - j0.5$	$\pm 0.9848 - j0.1736$	$\pm\frac{\sqrt{3}}{2} \pm j0.5$
		$\pm 0.342 + j0.9397$	$\pm 0.5 \pm j\frac{\sqrt{3}}{2}$
		$\pm 0.6428 - j0.766$	

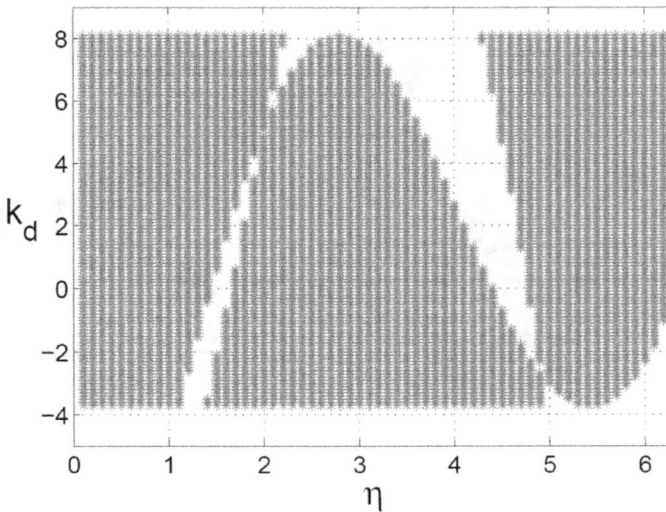

Fig. 2.3 K_d vs η as computed by applying the first procedure to the $PD^{1/3}$ controller $(K_i = 0, a = 0, b = 3, c = 1)$.

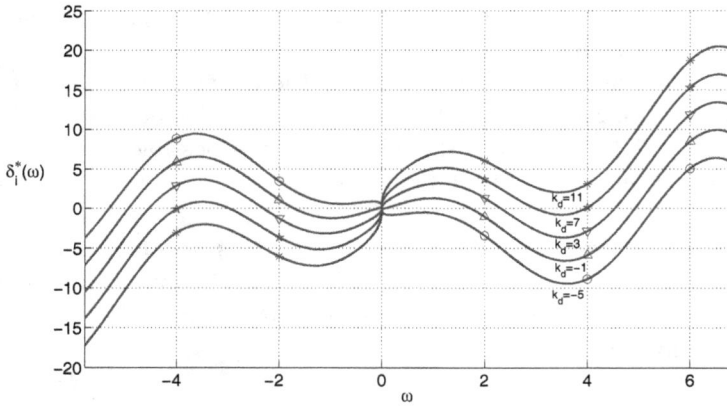

Fig. 2.4 Trend of $\delta_i^*(\omega)$ for different values of K_d for the $PD^{1/3}$ controller ($K_i = 0$, $a = 0, b = 3, c = 1$).

Then $\delta_i^*(\omega)$ has all real roots in the range $-3.7 < K_d < 8.1$, according the Pontryagin theorem.

The successive step is to determine the ranges of the values of K_p and K_d that fulfil the interlacing condition between the roots of $\delta_r^*(\omega)$ and $\delta_i^*(\omega)$. In order to verify this property the second developed procedure calculates the values of the real part $\delta_r^*(\omega)$ in the zeros of the imaginary part $\delta_i^*(\omega)$ according to the change of the parameters K_p and K_d.

By fixing $K_i = 0$ and $a = 0$ in the eqs. (2.15-2.18) for $\omega \neq 0$ the real part $\delta_r^*(\omega)$ can be rewritten as:

$$\delta_r^*(\omega) = k \Big[K_p + m(\omega) K_d + b(\omega) \Big] \qquad (2.28)$$

where

$$m(\omega) = \Big| \mathrm{Re}\big\{ (j)^{\frac{c}{b}} \big\} \Big| |\omega|^{\frac{c}{b}} \left(\frac{1}{L} \right)^{\frac{c}{b}}, \qquad (2.29)$$

and

$$b(\omega) = \frac{1}{k} \Big[\big(\cos \omega - \frac{T}{L} \omega \sin \omega \big) \Big] \qquad (2.30)$$

Since $\delta_i^*(\omega)$ is an odd function, it always has a root in $\omega = 0$. Thus, for $\omega = \omega_0 = 0$

$$\delta_r^*(\omega_0) = k K_p + 1 \qquad (2.31)$$

while for $\omega = \omega_j \neq 0$

$$\delta_r^*(\omega_j) = k\left[K_p + m(\omega_j)K_d + b(\omega_j)\right] \tag{2.32}$$

Thus, in order to verify the interlace property between the roots of $\delta_r^*(\omega)$ and $\delta_i^*(\omega)$ one must impose

$$\delta_r^*(\omega_0) > 0 \Rightarrow kK_p + 1 > 0 \Rightarrow K_p > -\frac{1}{k} \tag{2.33}$$

and

$$(-1)^j \delta_r^*(\omega_j) > 0 \tag{2.34}$$
$$\Rightarrow (-1)^j m(\omega_j)K_d + (-1)^j n(\omega_j)K_i + (-1)^j b(\omega_j) > (-1)^{j+1} K_p$$

taking into account that $k = 1$.

The conditions $\delta_r^*(\omega_0) > 0$ and $(-1)^j \delta_r^*(\omega_j) > 0$ have been selected rather than $\delta_r^*(\omega_0) < 0$ and $(-1)^j \delta_r^*(\omega_j) < 0$ because from the third developed procedure in order to fulfil the second condition of the Hermite-Biehler theorem the condition $K_p > -1/k$ is obtained, as it will be shown successively.

Thus, the range of the values of the parameters K_p and K_d that fulfil both the conditions of the Hermite-Biehler theorem are shown in Fig. 2.5.

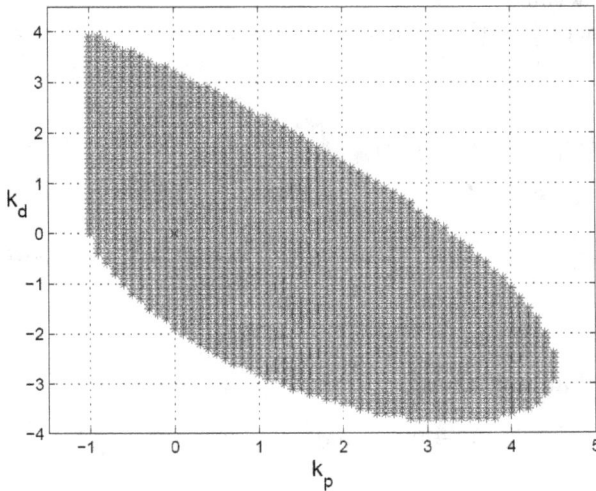

Fig. 2.5   Stability region of the $PD^{1/3}$ controller ($K_i = 0, a = 0, b = 3, c = 1$).

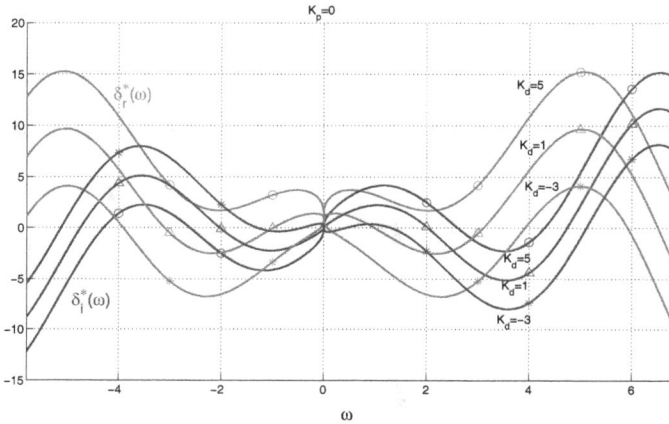

Fig. 2.6   Trend of $\delta_r^*(\omega)$ and $\delta_i^*(\omega)$ for different values of K_d for the $PD^{1/3}$ controller ($K_p = 0$ $K_i = 0$, $a = 0, b = 3, c = 1$).

In Fig. 2.6 it is possible to note that by fixing the value of K_p, equal to 0 for example, out of the range $-1.9 < K_d < 3.2$ the interlacing condition between the roots of the real and imaginary part of the characteristic equation is not fulfilled.

However, the fact that the second condition of the Hermite-Biehler theorem for a fractional PD is always verified for $K_p > -1/k$ still needs to be shown. In fact after that, the derivatives with respect to ω of $\delta_r^*(\omega)$ and $\delta_i^*(\omega)$ in eqs. (2.24 - 2.25) are calculated:

$$\delta'^*_r(\omega) = kK_d\left|\text{Re}\left\{(j)^{\frac{c}{b}}\right\}\right|\frac{c}{b}|\omega|^{\frac{c}{b}-1}\left(\frac{1}{L}\right)^{\frac{c}{b}} - \sin(\omega) - \frac{T}{L}\left(\sin\omega + \omega\cos\omega\right)$$
(2.35)

$$\delta'^*_i(\omega) = kK_d\left|\text{Im}\left\{(j)^{\frac{c}{b}}\right\}\right|\frac{c}{b}|\omega|^{\frac{c}{b}-1}\left(\frac{1}{L}\right)^{\frac{c}{b}}\text{sign}(\omega) + \cos(\omega)$$
$$+ \frac{T}{L}\left(\cos\omega - \omega\sin\omega\right)$$
(2.36)

and evaluated at $\omega = \overline{\omega} = 0$:

$$\delta'^*_r(0) = 0 \quad \text{and} \quad \delta'^*_i(0) = 1 + \frac{T}{L}$$
(2.37)

as well as $\delta_r^*(\omega)$ and $\delta_i^*(\omega)$ are evaluated at $\overline{\omega} = 0$:

$$\delta_r^*(0) = kK_p + 1 \quad \text{and} \quad \delta_i^*(0) = 0$$
(2.38)

Table 2.2 Complex solutions of $(j)^{\frac{2}{3}}$

$PD^{2/3}$	$PD^{4/6}$	$PD^{6/9}$	$PD^{8/12}$
-1	± 1	-1	± 1
$0.5 \pm j\frac{\sqrt{3}}{2}$	$\pm 0.5 \pm j\frac{\sqrt{3}}{2}$	$0.5 \pm j\frac{\sqrt{3}}{2}$	$\pm j$
		$-0.1736 \pm j0.9848$	$\pm 0.5 \pm j\frac{\sqrt{3}}{2}$
		$0.9397 \pm j0.342$	$\pm \frac{\sqrt{3}}{2} \pm j0.5$
		$-0.766 \pm j0.6428$	

Table 2.3 Complex solutions of $(j)^{\frac{1}{2}}$

$PD^{1/2}$	$PD^{2/4}$	$PD^{3/6}$	$PD^{4/8}$
$\frac{\sqrt{2}}{2} + j\frac{\sqrt{2}}{2}$	$\pm \frac{\sqrt{2}}{2} \pm j\frac{\sqrt{2}}{2}$	$\frac{\sqrt{2}}{2} + j\frac{\sqrt{2}}{2}$	± 1
$-\frac{\sqrt{2}}{2} - j\frac{\sqrt{2}}{2}$		$-\frac{\sqrt{2}}{2} - j\frac{\sqrt{2}}{2}$	$\pm j$
		$0.259 - j0.986$	$\pm \frac{\sqrt{2}}{2} \pm j\frac{\sqrt{2}}{2}$
		$-0.259 + j0.986$	
		$0.986 - j0.259$	
		$-0.986 + j0.259$	

the second condition of the Hermite-Biehler theorem $\delta'^{*}_i(\varpi)\delta^{*}_r(\varpi) - \delta^{*}_i(\varpi)\delta'^{*}_r(\varpi) > 0$ implies $K_p > -1/k$ because T and L are positive.

This last condition is always verified for any fractional PD independently by the derivation order.

Via the same developed procedures the stability areas of the fractional PD parameters with fractional order of the derivative action equal to 2/3 and 1/2 are calculated and shown in Fig. 2.7 and Fig. 2.8, taking into account the solutions of $(j)^{0.\overline{6}}$ and $(j)^{0.5}$, shown in Tab. 2.2 and Tab. 2.3, respectively.

As previously noted, the common complex solution with the smallest positive phase is selected in order to be replaced in the characteristic equation.

In order to test the established stability areas of the $PI^{\lambda}D^{\mu}$ parameters the step response of the closed-loop system in Fig. 2.2 is calculated in the boundaries of these areas.

In particular, in order to test the calculated stability region of the $PD^{0.\overline{3}}$ parameters, shown in Fig. 2.5, the step responses calculated according to the parameters values corresponding to points near the boundaries, inside the stability region are presented in Fig. 2.9, while out of this region are presented in Fig. 2.10.

Similarly, in order to test the established stability region of the $PD^{0.\overline{6}}$ and $PD^{0.5}$ parameters, shown in Fig. 2.7 and Fig. 2.8 respectively, the step

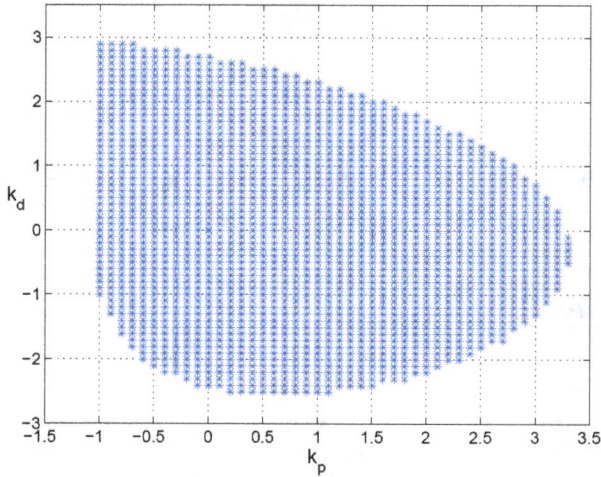

Fig. 2.7   Stability region of the $PD^{2/3}$ controller ($K_i = 0, a = 0, b = 3, c = 2$).

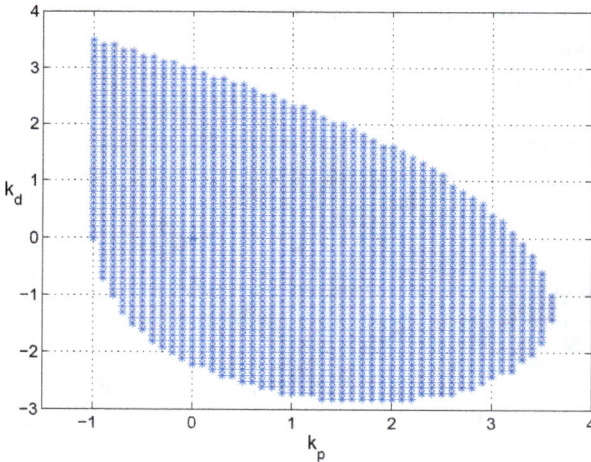

Fig. 2.8   Stability region of the $PD^{1/2}$ controller ($K_i = 0, a = 0, b = 2, c = 1$).

responses calculated according to the parameters values corresponding to points near the boundaries, inside the stability region and out of this region are presented in Fig. 2.11 and Fig. 2.12 for the $PD^{0.\bar{6}}$, while in Fig. 2.13 and Fig. 2.14 for the $PD^{0.5}$.

With the results in Figs. 2.9-2.14 it is possible to point out the accuracy of the established stability regions, obtained by the developed procedure, previously described, implementing the Pontryagin and Hermite-Biehler theorems.

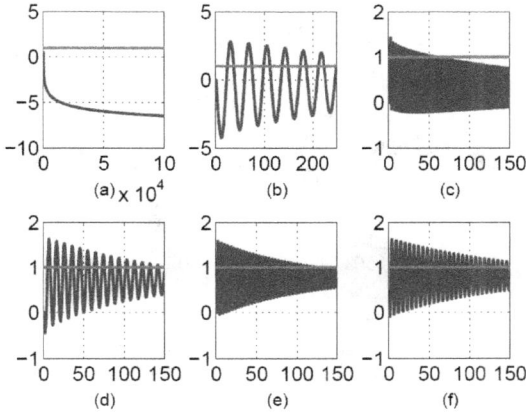

Fig. 2.9   Step responses of the closed-loop system where $C(s) = K_p + K_d s^{1/3}$ with (a) $K_p = -0.9, K_d = 2$, (b) $K_p = 0, K_d = -1.9$, (c) $K_p = 0, K_d = 3.2$, (d) $K_p = 3, K_d = -3.7$, (e) $K_p = 3, K_d = 0.3$, (f) $K_p = 4.5, K_d = -2.7$.

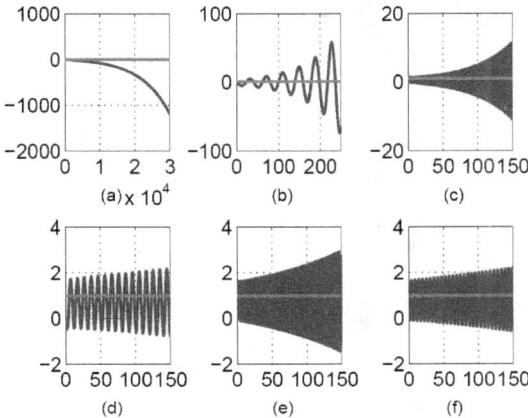

Fig. 2.10   Step responses of the closed-loop system where $C(s) = K_p + K_d s^{1/3}$ with (a) $K_p = -1.1, K_d = 2$, (b) $K_p = 0, K_d = -2$, (c) $K_p = 0, K_d = 3.3$, (d) $K_p = 3, K_d = -3.8$, (e) $K_p = 3, K_d = 0.4$, (f) $K_p = 4.6, K_d = -2.7$.

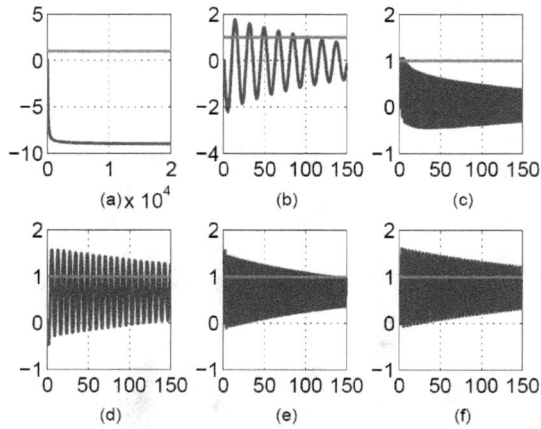

Fig. 2.11   Step responses of the closed-loop system where $C(s) = K_p + K_d s^{2/3}$ with (a) $K_p = -0.9, K_d = 1$, (b) $K_p = 0, K_d = -2.4$, (c) $K_p = 0, K_d = 2.7$, (d) $K_p = 2, K_d = -2.2$, (e) $K_p = 2, K_d = 1.7$, (f) $K_p = 3.3, K_d = -0.3$.

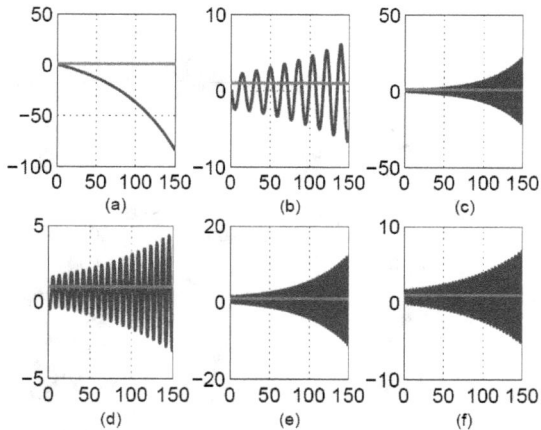

Fig. 2.12   Step responses of the closed-loop system where $C(s) = K_p + K_d s^{2/3}$ with (a) $K_p = -1.1, K_d = 1$, (b) $K_p = 0, K_d = -2.5$, (c) $K_p = 0, K_d = 2.8$, (d) $K_p = 2, K_d = -2.3$, (e) $K_p = 2, K_d = 1.8$, (f) $K_p = 3.4, K_d = -0.3$.

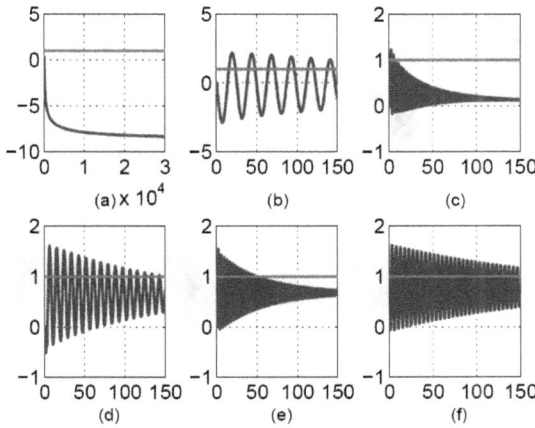

Fig. 2.13   Step responses of the closed-loop system where $C(s) = K_p + K_d s^{1/2}$ with (a) $K_p = -0.9, K_d = 2$, (b) $K_p = 0, K_d = -2.2$, (c) $K_p = 0, K_d = 2.9$, (d) $K_p = 2, K_d = -2.8$, (e) $K_p = 2, K_d = 1.5$, (f) $K_p = 3.6, K_d = -1.2$.

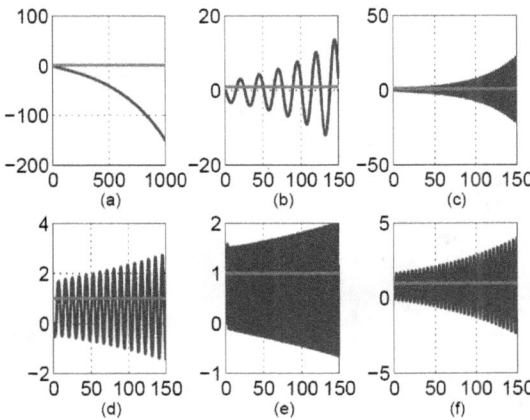

Fig. 2.14   Step responses of the closed-loop system where $C(s) = K_p + K_d s^{1/2}$ with (a) $K_p = -1.1, K_d = 2$, (b) $K_p = 0, K_d = -2.3$, (c) $K_p = 0, K_d = 3.1$, (d) $K_p = 2, K_d = -2.9$, (e) $K_p = 2, K_d = 1.6$, (f) $K_p = 3.7, K_d = -1.2$.

Chapter 3

Fractional Order Chaotic Systems

A brief introduction on the fractional order chaotic systems will be given in this chapter. In such a system the order of derivatives in its state-space representation is a fractional one. A survey of several well-known integer and fractional order chaotic systems will be presented.

3.1 Introduction

It is well-known that chaos cannot occur in continuous systems of having a total order less than three. This assertion is based upon the usual concepts of order, such as the number of states in a system or the total number of separate differentiations or integrations in the system. The model of a system can be rearranged in three single differential equations, where the equations contain the non-integer (fractional) order derivatives. The total order of the system is changed from 3 to the sum of each particular order. To put this fact into context, one can consider the fractional-order dynamical model of the system. In such cases the chaos was exhibited in a system with a total order less than three. The term "system order" should also be mentioned. The system order is not equal to the number of differential equations if one considers the fractional differential equations. The system order is equal to the highest derivative of the fractional differential equation of the mathematical model.

Generally, the following incommensurate fractional order nonlinear system is considered in the form:

$$
\begin{aligned}
{}_0D_t^{q_i} x_i(t) &= f_i(x_1(t), x_2(t), \ldots, x_n(t), t) \\
x_i(0) &= c_i, \quad i = 1, 2, \ldots, n,
\end{aligned}
\tag{3.1}
$$

where c_i are initial conditions, or in its their vector representation:

$$D^q\mathbf{x} = \mathbf{f}(\mathbf{x}), \tag{3.2}$$

where $\mathbf{q} = [q_1, q_2, \ldots, q_n]^T$ for $0 < q_i < 2$, $(i = 1, 2, \ldots, n)$ and $\mathbf{x} \in \mathbb{R}^n$.

The equilibrium points of the system (3.2) are calculated via the following equation

$$\mathbf{f}(\mathbf{x}) = 0 \tag{3.3}$$

and assuming that $x^* = (x_1^*, x_2^*, \ldots, x_n^*)$ is an equilibrium point of system (3.2).

3.2   Concept of Chua's System

3.2.1   *Classical Chua's Oscillator*

Classical Chua's oscillator, which is shown in Fig. 3.1, is a simple electronic circuit that exhibits nonlinear dynamical phenomena such as bifurcation and chaos. This circuit can be described by the equations [Matsumoto (1984)]:

$$\frac{dV_1(t)}{dt} = \frac{1}{C_1}\left[G(V_2(t) - V_1(t)) - f(V_1(t))\right],$$

$$\frac{dV_2(t)}{dt} = \frac{1}{C_2}\left[G(V_1(t) - V_2(t)) + I(t)\right], \tag{3.4}$$

$$\frac{dI(t)}{dt} = \frac{1}{L_1}\left[-V_2(t) - R_L I(t)\right],$$

where $G = 1/R_2$, $I(t)$ is the current through the inductance L_1, $V_1(t)$ and $V_2(t)$ are the voltages over the capacitors C_1 and C_2, respectively, and $f(V_1(t))$ is the piecewise-linear $v-i$ characteristic of nonlinear resistor (NR) - Chua's diode, depicted in Fig. 3.2, which can be described as

$$I_{NR}(t) = f(V_1(t)) = G_b V_1(t)$$
$$+ \frac{1}{2}(G_a - G_b)(|V_1(t) + B_p| - |V_1(t) - B_p|) \tag{3.5}$$

with B_p being the breakpoint voltage of a diode, and $G_a < 0$ and $G_b < 0$ being some appropriate constants (slope of the piecewise linear resistance).

By defining

$$x = V_1/B_p, \quad y = V_2/B_p, \qquad z = I_L/B_p G,$$
$$\alpha = C_2/C_1, \quad \beta = C_2/(L_1 G^2), \gamma = C_2 R/(LG), \tag{3.6}$$
$$m_1 = G_b/G, \quad m_0 = G_a/G, \qquad \tau = tG/C_2$$

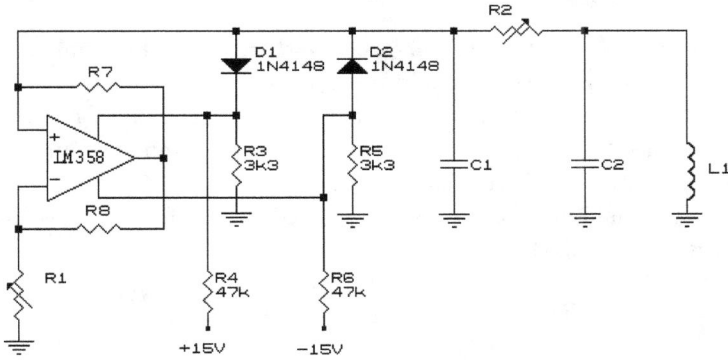

Fig. 3.1   Practical realization of Chua's circuit.

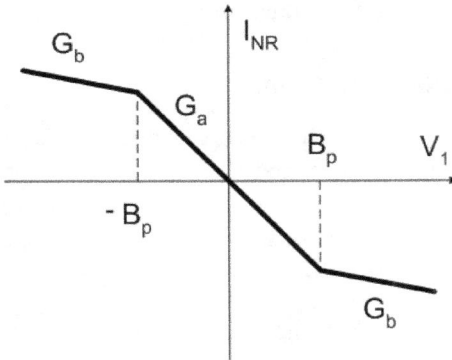

Fig. 3.2   Piecewise-linear $v - i$ characteristic of the nonlinear resistor.

one can transform (3.4) into the following corresponding dimensionless form of Chua's circuit [Deregel (1993)]:

$$\frac{dx(t)}{dt} = \alpha \left(y(t) - x(t) - f(x) \right),$$

$$\frac{dy(t)}{dt} = x(t) - y(t) + z(t), \tag{3.7}$$

$$\frac{dz(t)}{dt} = -\beta y(t) - \gamma z(t),$$

where

$$f(x) = m_1 x(t) + \frac{1}{2}(m_0 - m_1) \times \left(|x(t) + 1| - |x(t) - 1| \right) \tag{3.8}$$

and τ in transformation equations (3.6) is the dimensionless time.

3.2.2   *Chua-Hartley's Oscillator*

Given the techniques of fractional calculus, there are still a number of ways
in which the order of system could be amended.

The Chua - Hartley's system is different from the usual Chua's system
in that the piecewise-linear nonlinearity is replaced by an appropriate cubic
nonlinearity which yields very similar behaviour. Derivatives on the left side
of the differential equations are also replaced by the fractional derivatives
as follows [Hartley (1995)]:

$$_0D_t^q x(t) = \alpha \left(y(t) + \frac{x(t) - 2x^3(t)}{7} \right),$$

$$_0D_t^q y(t) = x(t) - y(t) + z(t), \tag{3.9}$$

$$_0D_t^q z(t) = -\beta y(t) = -\frac{100}{7} y(t),$$

where $q \leq 1$, $q \in \mathrm{R}$ is the fractional order of derivatives.

3.2.3   *Chua-Podlubny's Oscillator*

This system uses an approach where the order of any of the three con-
stitutive equations can be changed (3.4) so that the total order gives the
desired value. In Chua-Podlubny's case, the first differentiation is replaced
by fractional differentiation of order $q < 1$, $q \in \mathrm{R}$ in the equation one. The
final dimensionless equations of the system are [Podlubny (1999a)]:

$$_0D_t^q x(t) = \alpha \, _0D_t^{q-1} \left(y(t) - x(t) \right) - \frac{2\alpha}{7} \left(4x(t) - x^3(t) \right),$$

$$\frac{dy(t)}{dt} = x(t) - y(t) + z(t), \tag{3.10}$$

$$\frac{dz(t)}{dt} = -\frac{100}{7} y(t) = -\beta y(t),$$

where $\alpha = C_2/C_1$ and $\beta = C_2 R_2^2 / L_1$.

3.2.4   *New Fractional-Order Chua's Oscillator*

There are a large number of electric and magnetic phenomena where fractional calculus can be used [Westerlund (2002)]. Here only two of them will
be considered - capacitor and inductance coil behaviours.

The circuit behaviour can be described by three fractional differential
equations with various orders. By applying the Kirchhoff laws for two
current nodes and one voltage loop and relations $I(t) = C\frac{d^\alpha V(t)}{dt^\alpha}$, and

$V(t) = L\frac{d^\alpha I(t)}{dt^\alpha}$ into the circuit depicted in Fig. 3.1, the following mathematical model of the circuit for state variables $V_1(t)$, $V_2(t)$ and $I(t)$, is obtained:

$$C_1\,_0D_t^{q_1}V_1(t) + I_{NR}(t) = \frac{V_2(t) - V_1(t)}{R_2},$$

$$C_2\,_0D_t^{q_2}V_2(t) - I(t) = \frac{V_1(t) - V_2(t)}{R_2}, \qquad (3.11)$$

$$L_1\,_0D_t^{q_3}I(t) + V_2(t) + R_LI(t) = 0.$$

The equations (3.11) can be rewritten in the following form:

$$_0D_t^{q_1}V_1(t) = \frac{1}{C_1R_2}[V_2(t) - V_1(t)] - \frac{f(V_1(t))}{C_1},$$

$$_0D_t^{q_2}V_2(t) = \frac{1}{C_2R_2}[V_1(t) - V_2(t)] + \frac{I(t)}{C_2}, \qquad (3.12)$$

$$_0D_t^{q_3}I(t) = \frac{1}{L_1}[-V_2(t) - R_LI(t)],$$

where V_1 is a voltage across the capacitor C_1, V_2 is a voltage across the capacitor C_2, I is a current through the inductance L_1, q_1 is a real order of the capacitor C_1, q_2 is a real order of the capacitor C_2, q_3 is a real order of the inductor L_1, $f(V_1)$ is a is the piecewise linear $v - i$ characteristic of nonlinear Chua's diode, which can be described by (3.5).

By using the transformation (3.6), one can rewrite the equations (3.12) in the following dimensionless form [Petráš (2008)]:

$$_0D_t^{q_1}x(t) = \alpha\left(y(t) - x(t) - f(x)\right),$$

$$_0D_t^{q_2}y(t) = x(t) - y(t) + z(t), \qquad (3.13)$$

$$_0D_t^{q_3}z(t) = -\beta y(t) - \gamma z(t),$$

where $f(x)$ is the piecewise-linear nonlinearity (3.8).

3.2.4.1 *Experimental Measurements*

The classical Chua's oscillator can be realized by electrical elements according to the scheme shown in Fig. 3.1 , which is a simple electronic circuit [Kennedy (1992)] that exhibits nonlinear dynamical phenomena such as bifurcation and chaos. Chua's diode (3.5) – negative impedance converter – nonlinear resistor – was realized by operating amplifier LM 358 and resistors R_1, R_7 and R_8 $(R_7 = R_8)$.

The following values of electrical elements were chosen for the experimental verification of Chua's system, depicted in Fig. 3.1 and described by equations (3.12) and (3.5):

$$C_1 = 4.71nF, C_2 = 48nF, L_1 = 4.64mH, \qquad\qquad (3.14)$$
$$R_L = 15.8\Omega, R_1 = 897\Omega, R_2 = 998\Omega, R_7 = R_8 = 393\Omega$$

The metalized paper capacitors C_1 and C_2 with the real order $q_1 = q_2 = 0.98$ are used and the real order of inductor $q_3 = 0.94$ [Westerlund (2002)] is assumed. The total order of the system is $\bar{q} = 2.90$.

The measured breakpoints of the non-linear characteristic (3.5) are:

$$-B_p = (-8.79V, 7.7mA), \quad B_p = (9.12V, -7.9mA).$$

Assuming the three-segment piecewise-linear voltage-current transfer characteristic having a negative impedance converter (3.5), we obtain the slope $G_a = -1/R_1 = -1.1148 \, mA/V$ for $R_7 = R_8$ and the slope G_b is calculated using the breakpoints B_p and it has the value $G_b = -0.8710 \, mA/V$.

The resistors R_3, R_4, R_5, R_6, and the diodes D_1 and D_2 generate the positive and negative half of the non-linearity.

Fig. 3.3 depicts the photo of the digital oscilloscope screen (Tektronix TDS1002, 60 Mhz). It is a real measurement of voltages $V_1 - V_2$ for the circuit presented in Fig. 3.1 with the parameters of the electrical components (3.14). The result shown in Fig. 3.3 is the double-scroll attractor of fractional order Chua's system described by the equations (3.12) and (3.5).

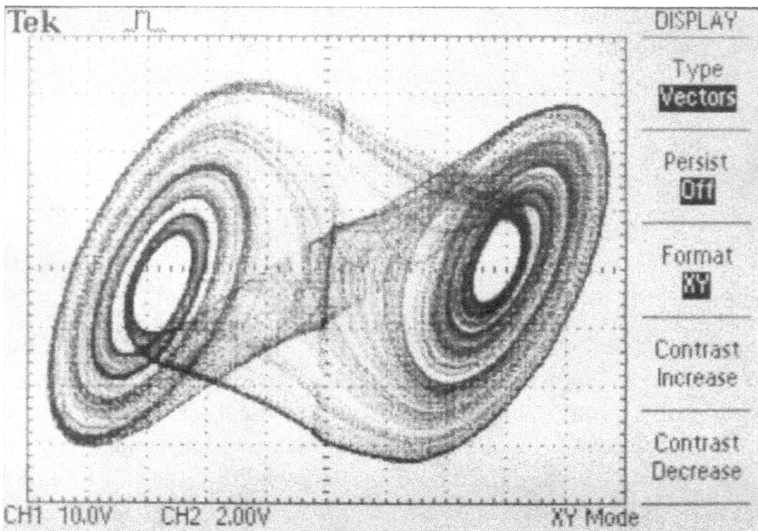

Fig. 3.3   Photo of oscilloscope screen: Strange attractor of the Chua's system (3.12).

3.2.4.2 *Simulation Results*

The relation derived from the Grünwald-Letnikov definition can be used for the numerical calculation of fractional-order derivation.

The similar and comparable results of Chua's fractional order system behaviour can be obtained by numerical simulation for time step $h = 0.001$, $L_m = 10$ (10000 values and coefficients from history).

Fig. 3.4 shows the double-scroll attractor of Chua's circuit (3.13) computed numerically for initial conditions $(x(0), y(0), z(0)) = (0.6, 0.1, -0.6)$ and the parts values (3.14).

3.3 Fractional-Order Van der Pol Oscillator

The Van der Pol oscillator (VPO) represents a nonlinear system with an interesting behaviour that arises naturally in several applications. It has been used in the study and the design of many models including biological phenomenas, such as the heartbeat, neurons, acoustic models, radiation of mobile phones, and as a model of electrical oscillators.

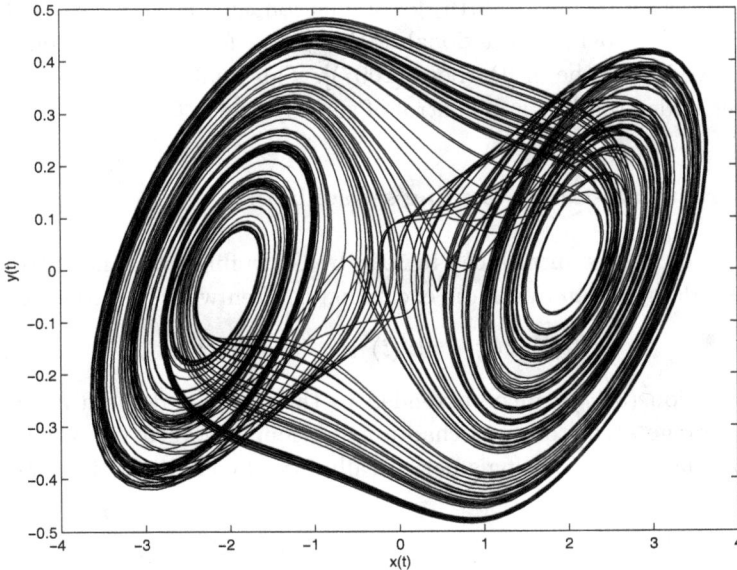

Fig. 3.4 Strange attractor of the fractional-order Chua's system (3.13) with total order $\bar{q} = 2.90$ for the parameters: $\alpha = 10.1911$, $\beta = 10.3035$, $\gamma = 0.1631$, $q_1 = q_2 = 0.98$, $q_3 = 0.94$, $m_0 = -1.1126$ and $m_1 = -0.8692$.

The VPO model was used by Van der Pol in 1920 to study oscillations in vacuum tube circuits. In the standard form, it is given by a nonlinear differential equation such as:

$$y''(t) + \epsilon(y(t)^2 - 1)y'(t) + y(t) = 0, \qquad (3.15)$$

where ϵ is the control parameter.

A modified version of the classical VPO was proposed by the fractional derivative of order q in a state space formulation of equation (3.15). It has the following form [Barbosa (2007)]:

$$_0D_t^q y_1(t) = y_2$$

$$\frac{dy_2}{dt} = -y_1 - \epsilon(y_1^2 - 1)y_2 \qquad (3.16)$$

In this section $0 < q < 1$ and $\epsilon > 0$ is taken into consideration. The resulting fractional-order Van der Pol oscillator (FrVPO) is reduced to the classical VPO when $q = 1$. The total system order is changed from the integer value 2 to the fractional value $1 + q < 2$.

Fig. 3.5 depicts the limit cycle in the phase plane of the Van der Pol fractional-order oscillator (3.16) for simulation time 30 sec. Detailed analysis of the Van der Pol fractional-order system for various system orders $1 + q$ have been done in [Barbosa (2007)]. This analysis may be useful for a better understanding and control of such systems.

3.4   Fractional-Order Duffing's Oscillator

Duffing's oscillator, introduced in 1918 by G. Duffing, with negative linear stiffness, damping and periodic excitation is often written in the form

$$x''(t) - x(t) + \alpha x'(t) + x^3(t) = \delta \cos(\omega t). \qquad (3.17)$$

The equation (3.17) can be extended to the complex domain in order to study strange attractors and chaotic bahaviour of forced vibrations of industrial machinery. Duffing's periodically forced complex oscillators of the form

$$z''(t) - z(t) + \alpha z'(t) + \epsilon z|z^2(t)| = \gamma' \cos(\omega t), \qquad (3.18)$$

where $\gamma' = \sqrt{2\gamma}\exp(i\pi/4)$, γ, α, ω are positive parameters, $z = x + iy$ is a complex function. Equation (3.18) can be reduced to the famous Duffing's oscillator (3.17) when $z = x, (y = 0)$ and $\epsilon = 1$. When $z = x + iy$ is

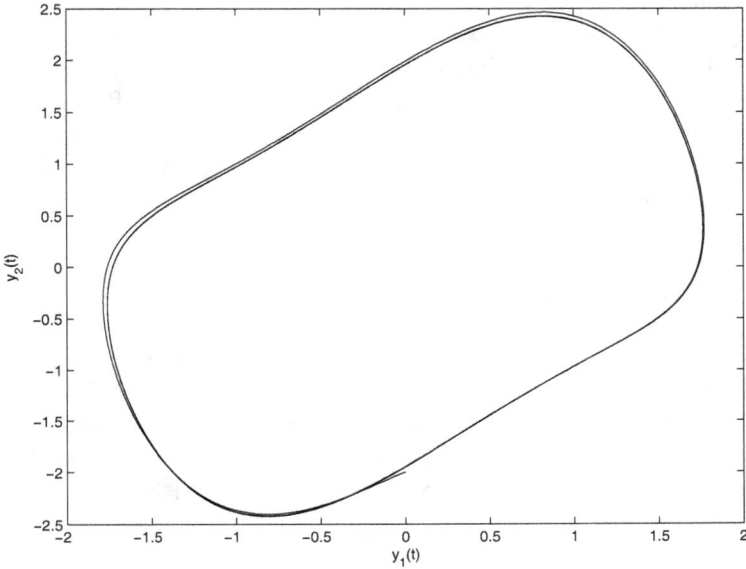

Fig. 3.5 Phase plane (y_1, y_2) plot (limit cycle) for FrVPO with fractional-order $q = 0.9$ and parameter $\epsilon = 1$. Initial conditions were: $\bar{y}_0 = [0, -2]$.

substituted by equation (3.18), this results in a system of two coupled nonlinear second-order differential equations [Gao (2005)]:

$$x''(t) - x(t) + \alpha x'(t) + \epsilon x(t)(x^2(t) + y^2(t)) = \gamma \cos(\omega t),$$
$$y''(t) - y(t) + \alpha y'(t) + \epsilon y(t)(x^2(t) + y^2(t)) = \gamma \cos(\omega t). \qquad (3.19)$$

To get Duffing's fractional-order system, equation (3.17) can be rewritten as a system of the first-order autonomous differential equations in the form:

$$\frac{x(t)}{dt} = y(t),$$
$$\frac{y(t)}{dt} = x(t) - x^3(t) - \alpha y(t) + \delta \cos(\omega t) \qquad (3.20)$$

Here, the conventional derivatives in equations (3.20) are replaced by the fractional derivatives as follows:

$$_0D_t^{q_1} x(t) = y(t),$$
$$_0D_t^{q_2} y(t) = x(t) - x^3(t) - \alpha y(t) + \delta \cos(\omega t), \qquad (3.21)$$

where q_1, q_2 are two fractional orders and α, δ, ω are system parameters.

Fig. 3.6 depicts the chaotic attractor of Duffing's system (3.20) for the following parameters $\alpha = 0.15$, $\delta = 0.3$, $\omega = 1$ with initial conditions $(x(0), y(0)) = (0.21, 0.13)$ simulation time for 200 sec.

Fig. 3.7 depicts the double scroll attractor of Duffing's fractional order system (3.21) for the following parameters $\alpha = 0.15$, $\delta = 0.3$, $\omega = 1$, derivative orders $q_1 = 0.9, q_2 = 1.0$ with initial conditions $(x(0), y(0)) = (0.21, 0.13)$ for simulation time 200 sec.

3.5 Fractional-Order Lorenz's System

Lorenz's oscillator is a 3-dimensional dynamical system that exhibits chaotic flow. Lorenz's attractor was named after Edward N. Lorenz, who derived it from the simplified equations of convection rolls arising in the equations of the atmosphere in 1963. For the first time he used the term "butterfly effect", which in chaos theory means sensitive dependence on initial conditions. Lorenz titled a 1979 paper, "Predictability: Does the Flap of a Butterfly's Wings in Brazil Set Off a Tornado in Texas?" Small

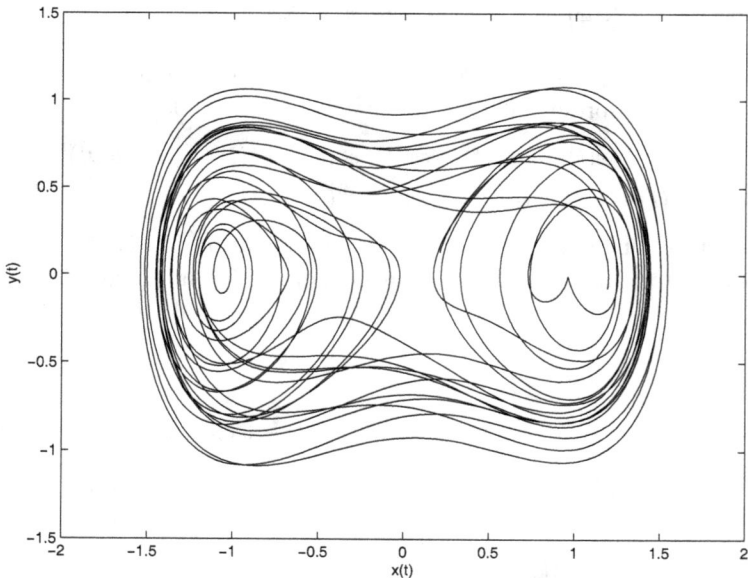

Fig. 3.6 Phase plane (x, y) plot for the Duffing's system (3.20) with parameters $\alpha = 0.15$, $\delta = 0.3$, $\omega = 1$, and initial conditions $(x(0), y(0)) = (0.21, 0.13)$.

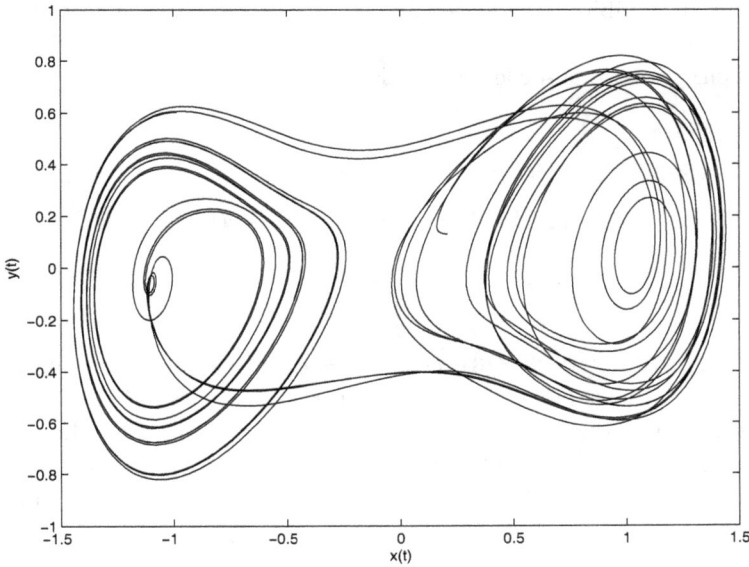

Fig. 3.7 Phase plane (x, y) plot (attractor) for the fractional order Duffing's system (3.21) with parameters $\alpha = 0.15$, $\delta = 0.3$, $\omega = 1$, derivative orders $q_1 = 0.9, q_2 = 1.0$, and initial conditions $(x(0), y(0)) = (0.21, 0.13)$.

variations of the initial condition of a dynamical system may produce large variations in the long term behaviour of the system. The phrase refers to the idea that a butterfly's wings might create tiny changes in the atmosphere that may ultimately alter the path of a tornado or delay, accelerate or even prevent the occurrence of a tornado in a certain location. The flapping wing represents a small change in the initial condition of the system, which causes a chain of events leading to large-scale alterations of events.

Lorenz's chaotic system is desribed by

$$\frac{dx(t)}{dt} = \sigma(y(t) - x(t)),$$

$$\frac{dy(t)}{dt} = x(t)(\rho - z(t)) - y(t), \qquad (3.22)$$

$$\frac{dz(t)}{dt} = x(t)y(t) - \beta z(t),$$

where σ is called the Prandtl number and ρ is called the Rayleigh number. All σ, ρ, $\beta > 0$, but usually $\sigma = 10$, $\beta = 8/3$ and ρ is varied.

The system exhibits chaotic behaviour for $\rho = 28$ and displays orbits for other values.

Lorenz's fractional-order system is described as (e.g. [Li (2007)]):

$$
\begin{aligned}
{}_0D_t^{q_1} x(t) &= \sigma(y(t) - x(t)), \\
{}_0D_t^{q_2} y(t) &= x(t)(\rho - z(t)) - y(t), \\
{}_0D_t^{q_3} z(t) &= x(t)y(t) - \beta z(t),
\end{aligned}
\tag{3.23}
$$

where $(\sigma, \rho, \beta) = (10, 28, 8/3)$, $q_1 = q_2 = q_3 = 0.993$, Lorenz's fractional order system (3.23) has a chaotic attractor.

Fig. 3.8 and Fig. 3.9 depict the simulation results of the Lorenz system (3.23) for the following parameters: $\sigma = 10, \rho = 28, \beta = 8/3$, orders $q_1 = q_2 = q_3 = 0.993$ and computational time 100 sec for time step $h = 0.005$.

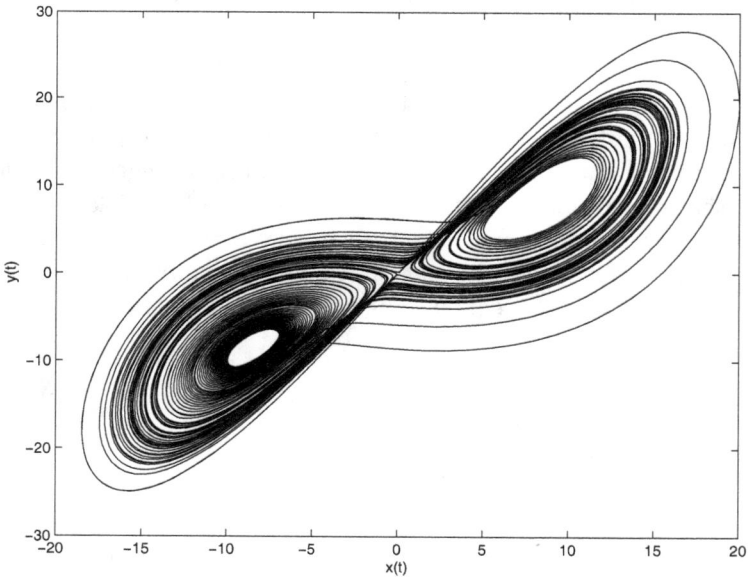

Fig. 3.8 Simulation result of the Lorenz's system (3.23) in $x - y$ plane for initial conditions $(x(0), y(0), z(0)) = (0.1, 0.1, 0.1)$.

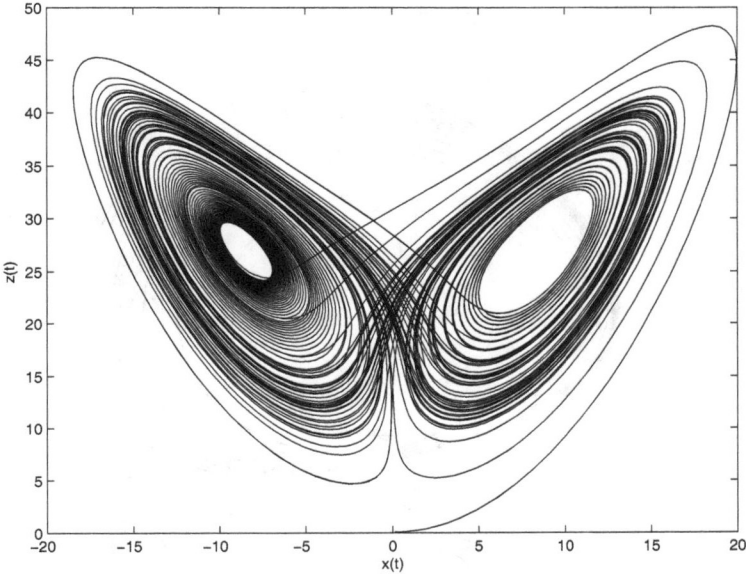

Fig. 3.9   Simulation result of the Lorenz's system (3.23) in $x - z$ plane for initial conditions $(x(0), y(0), z(0)) = (0.1, 0.1, 0.1)$.

3.6   Fractional-Order Genesio-Tesi System

The Genesio-Tesi system is described by

$$
\begin{aligned}
\frac{dx(t)}{dt} &= y(t), \\
\frac{dy(t)}{dt} &= z(t), \\
\frac{dz(t)}{dt} &= -\beta_1 x(t) - \beta_2 y(t) - \beta_3 z(t) + \beta_4 x^2(t),
\end{aligned}
\tag{3.24}
$$

where $\beta_1, \beta_2, \beta_3$ and β_4 are system parameters.

The Genesio-Tesi fractional-order system is defined as [Guo (2005)]:

$$
\begin{aligned}
{}_0 D_t^{q_1} x(t) &= y(t), \\
{}_0 D_t^{q_2} y(t) &= z(t), \\
{}_0 D_t^{q_3} z(t) &= -\beta_1 x(t) - \beta_2 y(t) - \beta_3 z(t) + \beta_4 x^2(t),
\end{aligned}
\tag{3.25}
$$

where $q \in [q_1, q_2, q_3]$ and $0 < q \leq 1$.

Fig. 3.10 depicts the simulation result of the Genesio-Tesi system (3.25) for the following parameters: $\beta_1 = 6.5$, $\beta_2 = 2.92$, $\beta_3 = 1.2$, $\beta_4 = 1.0$,

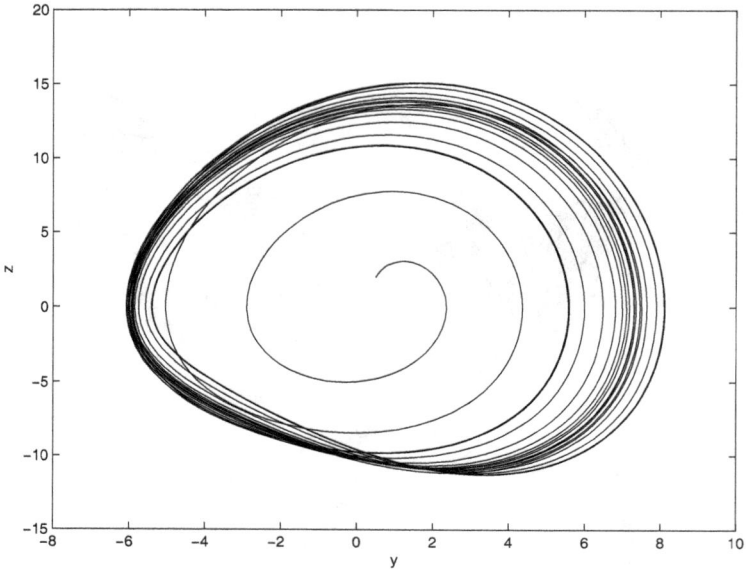

Fig. 3.10   Simulation result of the Genesio-Tesi system (3.25) in $y - z$ plane for initial conditions $(x(0), y(0), z(0)) = (-2, 0.5, 2)$.

orders $q_1 = 0.9$, $q_2 = 1.0$, $q_3 = 1.0$ and computational time 200 sec for time step $h = 0.005$.

3.7   Fractional-Order Lu's System

The so-called Lu's system is know as a bridge between Lorenz's system and the Chen's system. Its fractional version is described by the following equations [Deng (2005)]:

$$
\begin{aligned}
{}_0 D_t^{q_1} x(t) &= a(y(t) - x(t)), \\
{}_0 D_t^{q_2} y(t) &= -x(t)z(t) + cy(t), \\
{}_0 D_t^{q_3} z(t) &= x(t)y(t) - bz(t),
\end{aligned}
\tag{3.26}
$$

where $0 < q_1, q_2, q_3 \leq 1$, its order denoted by $q = (q_1, q_2, q_3)$ here, and a, b, c are system parameters.

Fig. 3.11 depicts the projection onto $x - y$ plane for derivative orders $q_1 = 0.985$, $q_2 = 0.990$, $q_3 = 0.980$ and parameters $a = 36$, $b = 3$,

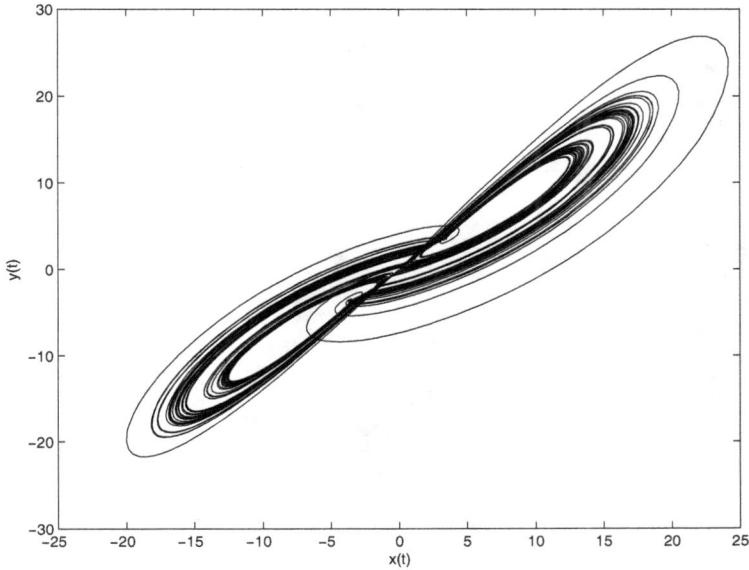

Fig. 3.11   Projection onto $x - y$ plane of Lu's fractional-order system (3.26) for parameters $a = 36, b = 3, c = 20$ and orders $q \in (0.985, 0.99, 0.98)$ for simulation time 60 sec.

$c = 20$ for simulation time 60 sec, and for the following initial conditions: $(x(0),\ y(0),\ z(0)) = (0.2,\ 0.5,\ 0.1)$.

3.8   Fractional-Order Rossler's System

Otto Rossler designed the Rossler's attractor in 1976, but the originally theoretical equations were later found to be useful in modeling equilibrium in chemical reactions. This attractor has only one manifold.

Let's consider a fractional-order generalization of the Rossler's system, where the conventional derivative is replaced by a fractional derivative, as follows [Li (2004)]:

$$
\begin{aligned}
{}_0D_t^q x(t) &= -(y(t) + z(t)), \\
{}_0D_t^q y(t) &= x(t) + ay(t), \\
{}_0D_t^q z(t) &= 0.2 + z(t)\,(x(t) - 10),
\end{aligned}
\tag{3.27}
$$

where system parameter a is allowed to be varied, and q is the fractional order.

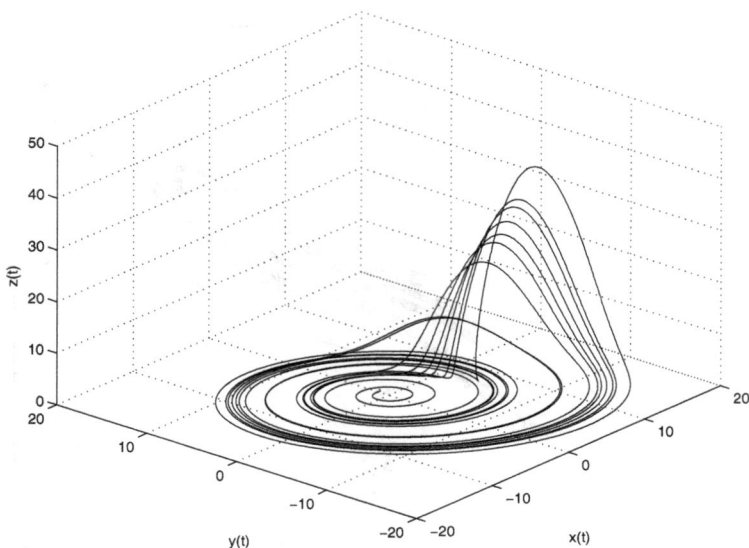

Fig. 3.12   Simulation result of Rossler's fractional-order system (3.27) in state space for parameter $a = 0.5$ and order $q = 0.9$ for simulation time 120 sec, for initial conditions $(x(0), y(0), z(0)) = (0.5, 1.5, 0.1)$.

Fig. 3.12 depicts the phase trajectory for derivative order $q = 0.9$ and parameter $a = 0.5$, for simulation time 120 sec, and for the initial conditions: $(x(0), y(0), z(0)) = (0.5, 1.5, 0.1)$.

3.9   Fractional-Order Newton-Leipnik System

The Newton-Leipnik system is described by the following non-linear differential equations:

$$\frac{dx(t)}{dt} = -ax(t) + y(t) + 10y(t)z(t),$$

$$\frac{dy(t)}{dt} = -x(t) - 0.4y(t) + 5x(t)z(t), \qquad (3.28)$$

$$\frac{dz(t)}{dt} = bz(t) - 5x(t)y(t),$$

where a and b are positive parameters. It is very interesting to note that the Newton-Leipnik system is a chaotic system with two strange attractors. When $(a, b) = (0.4, 0.175)$, with initial states $(0.349, 0, -0.16)$ and $(0.349, 0, -0.18)$, system (3.28) displays two strange attractors.

Here, the Newton-Leipnik fractional order system is considered, and the standard derivative is replaced by a fractional one as follows [Sheu (2008)]:

$$_0D_t^{q_1}x(t) = -ax(t) + y(t) + 10y(t)z(t),$$
$$_0D_t^{q_2}y(t) = -x(t) - 0.4y(t) + 5x(t)z(t), \qquad (3.29)$$
$$_0D_t^{q_3}z(t) = bz(t) - 5x(t)y(t),$$

where $0 < q_1, q_2, q_3 \leq 1$, its order denoted by $q = (q_1, q_2, q_3)$ here, and a, b are system parameters.

Fig. 3.13 and Fig. 3.14 depict the phase trajectories for derivative orders $q_1 = 0.99, q_2 = 0.99, q_3 = 0.99$ and parameters $a = 0.8, b = 0.175$ for simulation time 200 sec, and for the initial conditions: $(x(0), y(0), z(0)) = (0.19, 0.0, -0.18)$.

3.10 Fractional-Order Lotka-Volterra System

The Lotka-Volterra equations, also known as the predator-prey (or parasite-host) equations, are a pair of first order, non-linear, differential equations

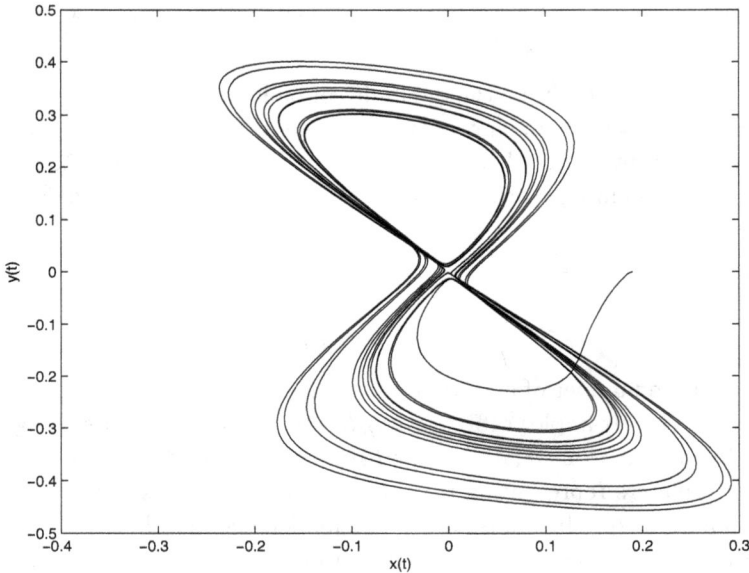

Fig. 3.13 Projection onto $x - y$ plane of the fractional-order Newton-Leipnik system (3.29) for parameters $a = 0.8, b = 0.175$ and orders $q \in (0.99, 0.99, 0.99)$ for simulation time 200 sec.

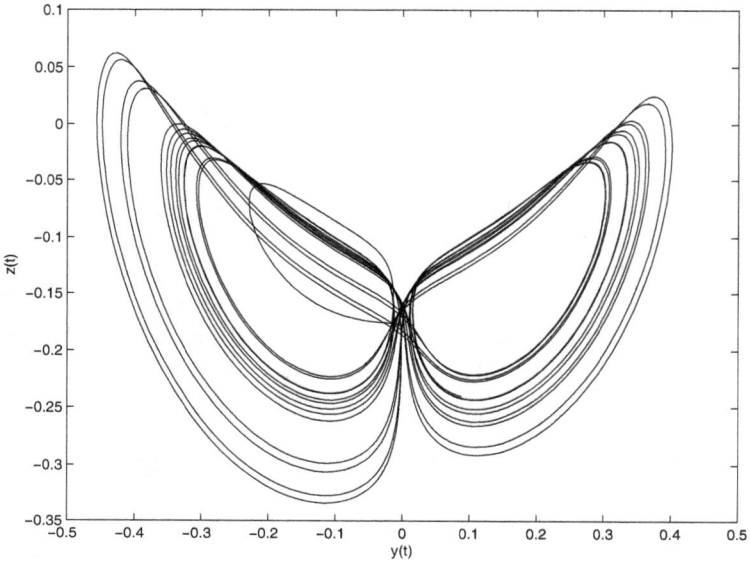

Fig. 3.14   Projection onto $y - z$ plane of the fractional-order Newton-Leipnik system (3.29) for parameters $a = 0.8, b = 0.175$ and orders $q \in (0.99, 0.99, 0.99)$ for simulation time 200 sec.

frequently used to describe the dynamics of biological systems in which two species interact, one being a predator and one its prey. They were proposed independently by Alfred J. Lotka in 1925 and Vito Volterra in 1926.

The classical integer-order model of the Lotka-Volterra system is defined as

$$\frac{dx}{dt} = x(t)(\alpha - \beta y(t))$$
$$\frac{dy}{dt} = -y(t)(\gamma - \delta x(t)),$$
(3.30)

where y is the number of a predator (for example, wolves); x is the number of its prey (for example, rabbits); dy/dt and dx/dt represents the growth of the two populations against time; t represents the time; and α, β, γ and δ are parameters representing the interaction of the two species.

Population equilibrium occurs in the model when neither of the population levels undergo any change, i.e. when both of the differential equations are equal to 0, we obtain

$$x(t)(\alpha - \beta y(t)) = 0$$
$$-y(t)(\gamma - \delta x(t)) = 0$$
(3.31)

When solved for x and y the above system of equations yields ($x = 0, y = 0$) and ($x = \lambda/\delta, y = \alpha/\beta$) hence there are two equilibria. The first solution effectively represents the extinction of both species. If both populations are at 0, then they will continue to be so indefinitely. The second solution represents a fixed point at which both populations sustain their current, non-zero numbers, and, in the simplified model, do so indefinitely. The levels of population at which this equilibrium is achieved depends on the chosen values of the parameters: $\alpha, \beta, \gamma, \delta$.

The Lotka-Volterra fractional-order (or fractional-order predator-prey model) system is described as [Ahmed (2007)]:

$$\begin{aligned}
{}_0D_t^q x(t) &= x(t)(\alpha - rx(t) - \beta y(t)) \\
{}_0D_t^q y(t) &= -y(t)(\gamma - \delta x(t)),
\end{aligned} \tag{3.32}$$

where $0 < q \le 1$, $x \ge 0$, $y \ge 0$ are prey and predator densities, respectively, and all constants r, α, β, γ and δ are positive.

Fig. 3.15 and Fig. 3.16 depict the phase trajectories for various derivative order $q = 1.0$ and $q = 0.8$, respectively, for simulation time 60 sec and for the initial conditions: $(x(0), y(0)) = (1, 2)$.

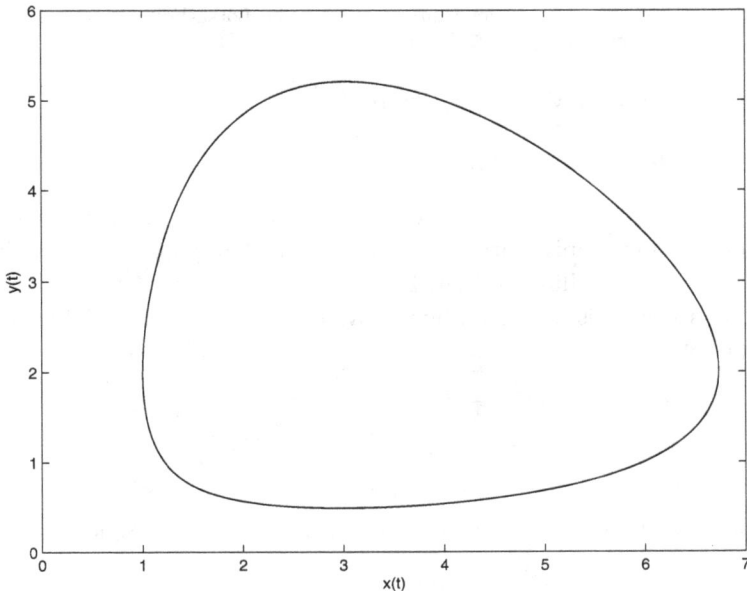

Fig. 3.15 Phase plane (x, y) plot (limit cycle) for the Lotka-Volterra system with order $q = 1.0$ and parameter $a = 2, b = 1, c = 3, d = 1, r = 0$.

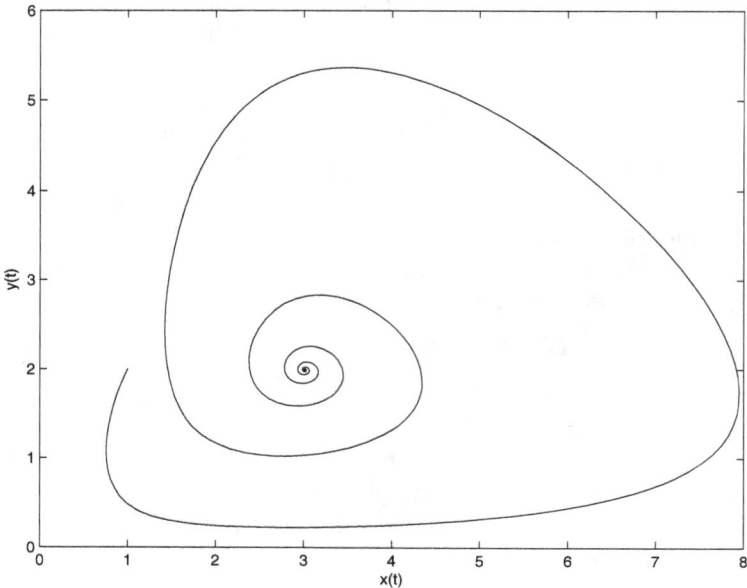

Fig. 3.16   Phase plane (x, y) plot (limit cycle) for the Lotka-Volterra with fractional-order $q = 0.9$ and parameter $a = 2, b = 1, c = 3, d = 1, r = 0$.

3.11   Concept of Volta's System

3.11.1   *Integer-Order Volta's System*

The system was discovered by Volta - a student at the Department of Physics, Genova University in 1984, whilst doing his thesis with the Prof. Antonio Borsellino and Dr. Francisco Fu Arcardi.

Volta's system is described by the system of state differential equations [Hao (1989)]:

$$\begin{aligned}
\dot{x}(t) &= -x(t) - 5y(t) - z(t)y(t), \\
\dot{y}(t) &= -y(t) - 85x(t) - x(t)z(t), \\
\dot{z}(t) &= 0.5z(t) + x(t)y(t) + 1.
\end{aligned} \tag{3.33}$$

Volta's system (3.33) can be generalised to the following form:

$$\begin{aligned}
\dot{x}(t) &= -x(t) - ay(t) - z(t)y(t), \\
\dot{y}(t) &= -y(t) - bx(t) - x(t)z(t), \\
\dot{z}(t) &= cz(t) + x(t)y(t) + 1.
\end{aligned} \tag{3.34}$$

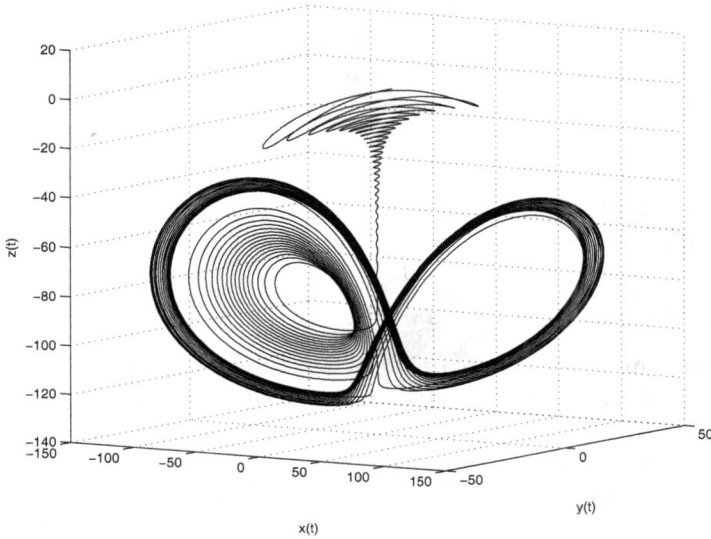

Fig. 3.17   Chaotic attractor of Volta's system (3.33) projected into 3D state space for initial conditions $(x(0), y(0), z(0)) = (8, 2, 1)$ and $T_{sim} = 20$ sec.

3.11.2   *Fractional-Order Volta's System*

Now, Volta's fractional-order system, where integer-order derivatives are replaced by fractional-order ones, will be considered. The mathematical description of the fractional-order chaotic system is expressed as [Petráš (2009a)]:

$$
\begin{aligned}
{}_0D_t^{q_1} x(t) &= -x(t) - ay(t) - z(t)y(t), \\
{}_0D_t^{q_2} y(t) &= -y(t) - bx(t) - x(t)z(t), \\
{}_0D_t^{q_3} z(t) &= cz(t) + x(t)y(t) + 1,
\end{aligned}
\tag{3.35}
$$

where q_1, q_2, and q_3 are the derivative orders. The total order of the system is $\bar{q} = (q_1, q_2, q_3)$.

3.11.2.1   *Case I: Commensurate Order*

When we assume the same orders of derivatives in state equations (3.35), i.e. $q_1 = q_2 = q_3 \equiv q$, we get a commensurate order system. According to the condition defined in [Tavazoei (2007b)], it is determined that the commensurate order q of derivatives has to be $q > 0.99$. It means, that for system parameters $(a, b, c) = (5, 85, 0.5)$, only for integer order $q = 1$, chaos can be observed. If we would like to go to the fractional (commensurate) order, we have to change the system parameters, e.g. for system parameters $(a, b, c) = (19, 11, 0.73)$, chaos can be observed if $q > 0.97$.

Fig. 3.18 shows the chaotic behaviour for fractional order system (3.35), where the system parameters are $(a, b, c) = (19, 11, 0.73)$, The commensurate order of the derivatives is $q = 0.98$, the initial conditions are $(x(0),\ y(0), z(0)) = (8, 2, 1)$ for simulation time $T_{sim} = 20$ sec and $h = 0.0005$.

3.11.2.2 *Case II: Incommensurate Order*

When we assume the different orders of derivatives in state equations (3.35), i.e. $q_1 \neq q_2 \neq q_3$, we get a general incommensurate order system. There is no exact condition to determine the orders to obtain the chaotic behaviour of the system. the following orders: $q_1 = 0.89$, $q_2 = 1.10$, and $q_3 = 0.91$ for system parameters $(a, b, c) = (5, 85, 0.5)$ were experimentally found.

Fig. 3.19 shows the chaotic behaviour for the fractional-order chaotic system (3.35), where the system parameters are $(a, b, c) = (5, 85, 0.5)$, the incommensurate orders of the derivatives are: $q_1 = 0.89$, $q_2 = 1.10$, and

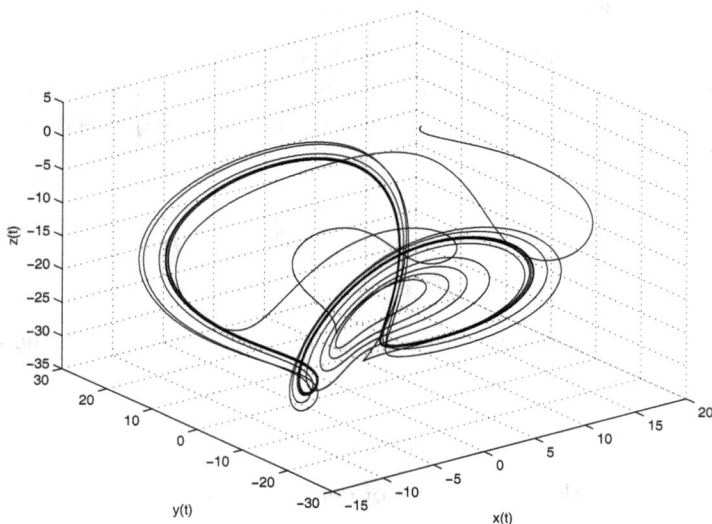

Fig. 3.18   Chaotic attractor of Volta's system (3.35) projected into 3D state space for initial conditions $(x(0), y(0), z(0)) = (8, 2, 1)$, parameters $(a, b, c) = (19, 11, 0.73)$, orders $(q_1, q_2, q_3) \equiv (q = 0.98)$ and $T_{sim} = 20$ sec.

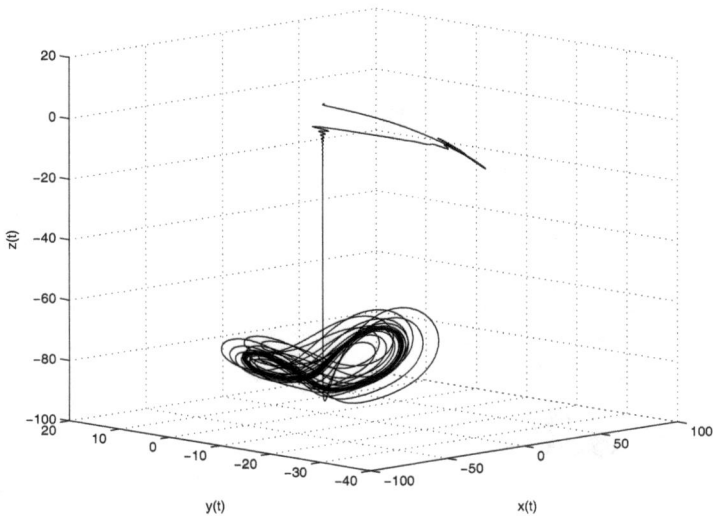

Fig. 3.19   Chaotic attractor of Volta's system (3.35) projected into 3D state space for initial conditions $(x(0), y(0), z(0)) = (8, 2, 1)$, parameters $(a, b, c) = (5, 85, 0.5)$, orders $(q_1, q_2, q_3) \equiv \bar{q} \in (0.89, 1.10, 0.91)$ and $T_{sim} = 20$ sec.

$q_3 = 0.91$, and the initial conditions are $(x(0), y(0), z(0)) = (8, 2, 1)$ for the simulation time $T_{sim} = 20$ sec and time step $h = 0.0005$. It can be envisaged that the behaviour of Volta's fractional system is chaotic and the double-scroll attractor [Tavazoei (2007b)] can be observed. The total order of the system is $\bar{q} = 2.9$.

Chapter 4

Field Programmable Gate Array Implementation

In this chapter the fundamental operator s^m, where m is a real number, is approximated via the binomial expansion of the backward difference. A hardware implementation of the differintegral operator is then proposed using the Field Programmable Gate Array (FPGA). Digital hardware implementation of a fractional-order differintegral operator requires a careful consideration of certain issues including system performance, hardware cost, and hardware speed. FPGA-based implementations are up to one hundred times faster than implementations based on microprocessors; this extra speed can be exploited to allow a higher performance in terms of digital approximations of fractional-order systems.

4.1 Numerical Fractional Integration

As explained in chapter 1, the Riemann-Liouville integral (see eq. (1.2)) is not always solvable in a closed form. For this reason, a numerical approximation is needed in order to allow an automatic evaluation of it. The aim of this study is to build a suitable numerical algorithm for any non integer order of integration. Obviously, the starting point is the RL definition, here rewritten as:

$$\frac{d^{-m}h(t)}{dt^{-m}} = \frac{1}{\Gamma(m)} \int_0^t (t-y)^{m-1}h(y)dy \qquad (4.1)$$

The interval $[0, t]$ is divided into k subintervals, $k+1$ being the number of samples of $f(t)$ which are indicated for simplicity, in the following way:

$$f_0 \equiv f(0)$$
$$f_1 \equiv f(\tfrac{t}{k})$$
$$\cdots\cdots\cdots$$
$$f_j \equiv f(\tfrac{jt}{k})$$ (4.2)
$$\cdots\cdots\cdots$$
$$f_k \equiv f(t)$$

The integral extended to the whole interval $[0, t]$ may be decomposed into the sum of k integrals:

$$\frac{d^{-m}h(t)}{dt^{-m}} = \frac{1}{\Gamma(m)} \sum_{j=0}^{k-1} \int_{\frac{jt}{k}}^{\frac{jt+t}{k}} (t-y)^{m-1} f(\tau) d\tau \qquad (4.3)$$

Assuming f is constant inside each subinterval (for example, equal to the mean value of the two extreme samples), its value can be extracted from the integral:

$$\int_{\frac{jt}{k}}^{\frac{jt+t}{k}} (t-y)^{m-1} f(\tau) d\tau \approx \frac{\frac{f(jt)}{k} + \frac{f((j+1)t)}{k}}{2} \int_{\frac{jt}{k}}^{\frac{jt+t}{k}} (t-y)^{m-1} d\tau$$

$$= \frac{f_j + f_{j+1}}{2q} \left\{ \left[t - \frac{jt}{k} \right]^m - \left[t - \frac{(j+1)t}{k} \right]^m \right\} \qquad (4.4)$$

Finally, the following formula is obtained:

$$\frac{d^{-m}f(0)}{dt^{-m}} = f(0)$$
$$\frac{d^{-m}f(k)}{dt^{-m}} = \frac{T^m}{\Gamma(1+m)} \sum_{j=0}^{k-1} \frac{f_j + f_{j+1}}{2} \left[(k-j)^m - (k-j-1)^m \right] \qquad (4.5)$$

where T is the sample time t/k. It must be noted that the calculation of every value requires all the preceding samples. In cases when a lot of steps are necessary, this fact may lead to a long computation time. However, the weight of the oldest samples is very small and for this reason the sum in (4.5) can be limited to the most recent ones, as will be demonstrated in section 4.3.

4.2   Grünwald-Letnikov Fractional Derivatives

This section is dedicated to the description of an approach towards the unification of two notions, which are usually presented separately in classical analysis: derivative of integer order n and $n-$fold integrals. As will be shown below, these notions are closer to each other than is usually assumed.

A continuous function $y = f(t)$ will now be considered. According to the well-known definition, the first-order derivative of the function $f(t)$ is defined as

$$f'(t) = \frac{df}{dt} = \lim_{h \to 0} \frac{f(t) - f(t-h)}{h} \tag{4.6}$$

Applying this definition twice gives the second-order derivative:

$$f''(t) = \frac{d^2 f}{dt^2} = \lim_{h \to 0} \frac{f'(t) - f'(t-h)}{h}$$

$$= \lim_{h \to 0} \frac{1}{h} \left\{ \frac{f(t) - f(t-h)}{h} - \frac{f(t-h) - f(t-2h)}{h} \right\}$$

$$= \lim_{h \to 0} \frac{f(t) - 2f(t-h) + f(t-2h)}{h^2} \tag{4.7}$$

Using (4.6) and (4.7) one obtains

$$f'''(t) = \frac{d^3 f}{dt^3} = \lim_{h \to 0} \frac{f(t) - 3f(t-h) + 3f(t-2h) - f(t-3h)}{h^3} \tag{4.8}$$

and, by induction,

$$f^{(n)}(t) = \frac{d^n f}{dt^n} = \lim_{h \to 0} \frac{1}{h^n} \sum_{r=0}^{n} (-1)^r \binom{n}{r} f(t - rh), \tag{4.9}$$

where

$$\binom{n}{r} = \frac{n(n-1)(n-2)\ldots(n-r+1)}{r!} \tag{4.10}$$

is the usual notation for the binomial coefficients.

The following expression generalizing the fractions in (4.6)-(4.10) will now be considered:

$$f_h^{(p)}(t) = \frac{1}{h^p} \sum_{r=0}^{N} (-1)^r \binom{p}{r} f(t - rh), \tag{4.11}$$

where p is an arbitrary integer number; n is also integer, as above.

Obviously, for $p < n$

$$\lim_{h \to 0} f_h^{(p)}(t) = f^{(p)}(t) = \frac{d^p}{dt^p} \tag{4.12}$$

because in such a case, as follows from (4.10), all the coefficients in the numerator after $\binom{p}{p}$ are equal to 0. Negative values of p will now be considered. For convenience,

$$\left[\begin{matrix} p \\ r \end{matrix} \right] = \frac{p(p+1)\ldots(p+r-1)}{r!} \tag{4.13}$$

then one has

$$\binom{-p}{r} = \frac{-p(-p-1)\dots(-p-r+1)}{r!} = (-1)^r \begin{bmatrix} p \\ r \end{bmatrix} \qquad (4.14)$$

and by replacing p in (4.11) with $-p$ one can write

$$f_h^{(-p)}(t) = \frac{1}{h^p} \sum_{r=0}^{n} \begin{bmatrix} p \\ r \end{bmatrix} f(t - rh), \qquad (4.15)$$

where p is a positive integer number.

If n is fixed, then $f_h^{(-p)}(t)$ tends to the uninteresting limit 0 as $h \to 0$. One can take $h = \frac{t-a}{n}$, where a is a real constant, and consider the limit value, either finite or infinite, of $f_h^{(-p)}(t)$, which will be denoted as

$$\lim_{\substack{h \to 0 \\ nh=t-a}} f_h^{(-p)}(t) = D_{[a,t]}^{-p} f(t) \qquad (4.16)$$

In fact, here $D_{[a,t]}^{-p} f(t)$ denotes a certain operation performed on the function $f(t)$; a and t are the *terminals* - the limits relating to this operation. Equation (4.16) suggests the following general expression:

$$D_{[a,t]}^{-p} f(t) = \lim_{\substack{h \to 0 \\ nh=t-a}} h^p \sum_{r=0}^{n} \begin{bmatrix} p \\ r \end{bmatrix} f(t - rh)$$

$$= \frac{1}{(p-1)!} \int_a^t (t - \tau)^{p-1} f(\tau) d\tau \qquad (4.17)$$

Now let us show that formula (4.17) is a representative of a p-fold integral. Integrating the relationship

$$\frac{d}{dt}\left(D_{[a,t]}^{-p} f(t) \right) = \frac{1}{(p-2)!} \int_a^t (t - \tau)^{p-2} f(\tau) d\tau = D_{[a,t]}^{-p+1} f(t) \qquad (4.18)$$

from a to t one obtains:

$$D_{[a,t]}^{-p} f(t) = \int_a^t \left(D_{[a,t]}^{-p+1} f(t) \right) dt \qquad (4.19)$$

$$D_{[a,t]}^{-p+1} f(t) = \int_a^t \left(D_{[a,t]}^{-p+2} f(t) \right) dt, \text{ etc.,} \qquad (4.20)$$

and therefore

$$D_{[a,t]}^{-p} f(t) = \int_a^t dt \int_a^t \left(D_{[a,t]}^{-p+2} f(t) \right) dt$$

$$= \int_a^t dt \int_a^t dt \left(D_{[a,t]}^{-p+3} f(t) \right) dt$$

$$= \underbrace{\int_a^t dt \int_a^t dt \dots \int_a^t f(t) dt}_{p \quad times} \qquad (4.21)$$

One can see that the derivative of an integer order n (4.9) and the p-fold integral (4.17) of the continuous function $f(t)$ are particular cases of the general expression

$$D_{[a,t]}^{-p} f(t) = \lim_{\substack{h \to 0 \\ nh=t-a}} h^p \sum_{r=0}^{n} (-1)^r \binom{p}{r} f(t - rh) \qquad (4.22)$$

which represents the derivative of order m if $p = m$ and the m-fold integral if $p = -m$.

This observation naturally leads to the idea of a generalization of the notions of differentiation and integration by allowing p in (4.22) to be an arbitrary real or a complex number. Here, attention will be restricted to the real values of p.

Thus, one can return to the Grünwald-Letnikov definition

$$D_{[a,t]}^{\alpha} f(t) = \lim_{h \to 0} \frac{\Delta_{[a,h]}^{\alpha} f(t)}{h^{\alpha}}, \quad \Delta_{[a,h]}^{\alpha} f(t) = \sum_{j=0}^{\left[\frac{t-a}{h}\right]} (-1)^j \binom{\alpha}{j} f(t - jh) \quad (4.23)$$

where $[x]$ means the integer part of x.

For a wide class of functions, important for applications, both definitions are equivalent. This allows one to use the Riemann-Liouville definition during problem formulation, and then turn to the Grünwald-Letnikov definition for obtaining the numerical solution.

Both definitions are equivalent for a wide class of functions which are important for applications. This allows the use of the Riemann-Liouville definition during problem formulation, and also makes it possible to use to the Grünwald-Letnikov definition to obtain the numerical solution.

4.3 The "Short-Memory" Principle

The following approximation, arising from the Grünwald-Letnikov definition, will be used:

$$D_{[a,t]}^{\alpha} f(t) \approx \Delta_{[a,h]}^{\alpha} f(t) \qquad (4.24)$$

Computations are performed using approximation (4.24).

For $t \gg a$ the number of addends in the fractional-derivative approximation (4.24) becomes enormously large. However, it follows from the expression for the coefficients in the Grünwald-Letnikov definition (4.23) that for large t the role of the "history" of the behaviour of the function $f(t)$ near the lower terminal (the "starting point") $t = a$ can be neglected under certain assumptions. These observations lead to the formulation of the "short-memory" principle, which means taking into account the behaviour of f(t) only in the "recent past", i.e. in the interval $[t - L, t]$, where L is the "memory length":

$$D_{[a,t]}^{\alpha} f(t) \approx D_{[t-L,t]}^{\alpha} f(t), \qquad (t > a + L) \tag{4.25}$$

In other words, according to the short-memory principle (4.25), the fractional derivative with the lower limit a is approximated by the fractional derivative with moving lower limit $t - L$. Due to this approximation, the number of addends in approximation (4.24) is always no more than $[L/h]$.

Of course, for this simplification a penalty is paid in the form of some inaccuracy. If $f(t) \leq M$ for $a \leq t \leq b$, which usually takes place in applications, then, the following estimate for the error introduced by the short-memory principle is calculated as follows:

$$\Delta(t) = |D_{[a,t]}^{\alpha} f(t) - D_{[t-L,t]}^{\alpha} f(t)| \leq \frac{M L^{-\alpha}}{|\Gamma(1 - \alpha)|}, (a + L \leq t \leq b) \tag{4.26}$$

This inequality can be used to determine the "memory length" L providing the required accuracy ϵ:

$$\Delta(t) \leq \epsilon, \quad (a + L \leq t \leq b), \quad if \quad L \geq \left(\frac{M}{\epsilon |\Gamma(1 - \alpha)|} \right)^{1/\alpha} \tag{4.27}$$

4.4 FPGA Hardware Implementation

4.4.1 *FPGA Introduction*

Fractional order systems have been studied extensively over the past two decades, and much progress has been made in the theory and analysis of these systems. However, little has been done in the area of hardware realization of fractional-order systems. In general, the implementation of fractional-order systems requires that they are approximated as high order rational systems. As a result, they are difficult to translate into hardware. The contribution of this work lies in providing a practical and efficient way to implement a fractional-order system. The implementation technique exploits the parallel structure and versatility of Field Programmable Gate Arrays (FPGAs) in order to yield a high performance and yet a low cost implementation. This work is the first step towards the development of a design flow to overcome the existing barrier between software-based simulations of fractional-order systems and real-time hardware solutions.

Digital hardware designers have a number of different computational platform options when implementing digital signal processing functions of the sort needed to approximate fractional-order systems. Historically, microprocessors and digital signal processors (DSPs) have dominated low-rate applications for which it is not crucial to save space and power. However, recent advances in technology and in the availability of system-level design tools from vendors have led to a rise in the popularity of using field programmable gate array (PPGAs) as a computational platform for digital signal processing applications [Caponetto (2007a)]. Field programmable gate arrays (FPGAs) are general-purpose integrated circuits that consist of tens of thousands of programmable logic cells interconnected by wires and programmable switches. The main advantage of an FPGA over a microprocessor or a DSP is its versatile, highly parallel structure. Since the programmable hardware can be tailored to implement the computation at hand in a maximally parallel way, it can outperform microprocessors and DSPs, which must run the computation serially on general purpose hardware. When combined with modern FPGA system clock rates in the hundreds of MHz, an FPGA-based implementation can be faster than a DSP processor by a ratio of 100 to 1. Overall, an FPGA has a computational power similar to that of an application specific integrated circuit (ASIC). However, unlike an ASIC, an FPGA is reconfigurable and has a low nonrecurring engineering cost and a short time-to-market.

Traditionally, implementation of high-order systems demands the use of floating-point computations. Use of the floating-point is more accurate and saves time in the early design phase, but makes for slower, more expensive hardware in production. In recent years, fixed-point operations have been widely used for hardware implementations in order to save costs and increase speed. The disadvantage of the fixed-point is that the hardware design requires a lot of careful consideration of the precisions required, with specific planning for each individual application. Thus, the hardware designer needs specific training in fixed-point considerations in order to develop a successful implementation.

4.4.2 *Remarks on the Fractional Differintegral Operator*

The fractional differintegral operator s^α ($\alpha \in R$) is the fundamental building block of any fractional-order system. To realize the fractional differintegral operator, generating function $s = \omega(z^{-1})$ with the z-transform of backward difference rule $\omega(z^{-1}) = (1 - z^{-1})/T$ is used, where T is the sampling time.

From the general binomial series expansion

$$(1 - x)^n = \sum_{j=0}^{\infty} (-1)^j \frac{n(n-1)(n-2)\ldots(n-j+1)}{j!} x^j \qquad (4.28)$$

one has

$$(w(z^{-1}))^\alpha = T^{-\alpha} \sum_{j=0}^{\infty} (-1)^j \frac{\alpha(\alpha-1)(\alpha-2)\ldots(\alpha-j+1)}{j!} x^j$$

$$= T^{-\alpha} \sum_{j=0}^{\infty} (-1)^j \binom{\alpha}{j} z^{-j} \qquad (4.29)$$

Therefore,

$$D^\alpha \equiv \lim_{N\to\infty} T^{-\alpha} \sum_{j=0}^{N} (-1)^j \binom{\alpha}{j} z^{-j} \qquad (4.30)$$

where

$$\binom{\alpha}{j} = \frac{\Gamma(\alpha+1)}{\Gamma(j+1)\Gamma(\alpha-j+1)} \qquad \text{and} \qquad \Gamma(n) = (n-1)! \qquad (4.31)$$

It should be noticed that this binomial series is similar to another discrete-time approximation based on the Grünwald-Letnikov definition, (see eq. (4.23)). Given by the expression

$$D^\alpha \equiv \lim_{N\to\infty} T^{-\alpha} \sum_{j=0}^{N} (-1)^j \binom{\alpha}{j} f(t - jT), \qquad (4.32)$$

it generates the same binomial coefficients $b_j = (-1)^j \binom{\alpha}{j}$.

In our algorithm we let

$$b_j = (-1)^j \binom{\alpha}{j} = \left(1 - \frac{1+\alpha}{j}\right) b_{j-1} \qquad \text{with} \qquad b_0 = 1. \qquad (4.33)$$

The binomial series expansion of the fractional differintegral operator has an infinite number of terms. The more terms are used in the approximation, the more accurately it will represents the original operator according to the "Short-Memory" Principle. From a practical standpoint, determination of an appropriate order for the approximation must consider a trade-off between functionality and hardware cost.

4.4.3 FPGA Implementation of the Fractional Differintegral Operator

In this section a hardware implementation of the formula (4.32) is explained. The hardware environment used for this application is the FPGA, which is able to guarantee high order digital approximation and high digital processing speed, as introduced in section 4.4.1.

The hardware block realizing the acquisition and the memorization of the input data samples is shown in Fig. 4.1.

The State-Machine manages the timing for the acquisition of the input signals, converted in digital form, and the timing for the memorization of the input signal samples.

The State-Machine, shown in Fig. 4.2, is characterized by three states. The State $S0$ is the reset state where the initialization of the variables is realized. In the State $S1$ the read phase is performed during which the input signal samples are acquired, while in State $S2$ the computation of the coefficients represented in eq. (4.33) is realized.

According to the control signals which manage the transitions from one state to another, the State Machine acquires the data and stores it in a Memory, shown in Fig. 4.1.

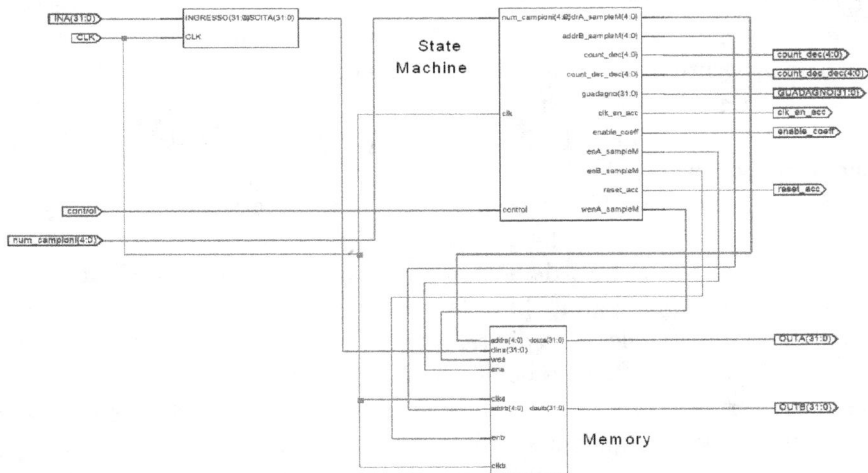

Fig. 4.1 Hardware block realizing the acquisition and the memorization of the input data samples.

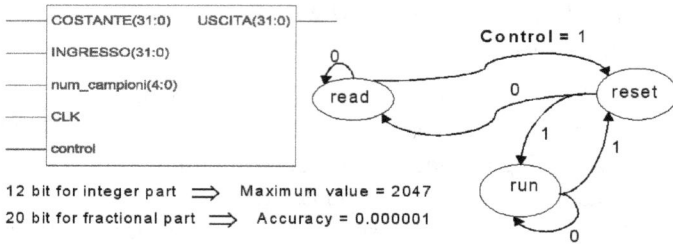

Fig. 4.2   State-Machine managing the acquisition and the memorization of the input signal samples.

The dimensions of the memory are defined according to the data coding and the accuracy, selected according to the "Short-Memory" Principle. In fact, the *length of the memory* is defined according to the coding, i.e. the number of bits used to represent an input data sample. Also, the amount of data stored in an FIFO memory, which is selected according to the "Short-Memory" Principle, determines the *depth of the memory*.

In this implementation the fixed-point coding is used to convert the analog data into a digital one. The data is converted into a bitstream of 32 bits, of which the first 12 bits are used to represent the integer part of the samples input data, while the remaining 20 bits are used to define the fractional value of the samples input data. This coding makes it possible to define a maximum value of about 2047 and an accuracy of about 10^{-6}.

The blocks that realize the computation of the coefficients in (4.33) are shown in Fig. 4.3. The data stored in the memory block are calculated according to the formula (4.33). These coefficients are stored, because of the recursive form of the formula (4.33).

The top hardware block scheme realizing the Fractional Differintegral Operator is shown in Fig. 4.4.

The input and output data of the realized FPGA Fractional Integral Operator have to be digital. In order to obtain this, i.e. the conversion of the data from an analog to a digital form and viceversa, two procedures in *Matlab* Environment have been developed. Thus the input and the output data, converted into analog form, are shown in Fig. 4.5 and 4.6 for an Integral Operator of fractional order 0.7 and 0.3, respectively.

The output signal is attenuated by a factor equal to $(2\pi f)^m = (2\pi 10)^{0.7} = 18.14$ and $(2\pi f)^m = (2\pi 10)^{0.3} = 3.46$, i.e. an input signal of amplitude equal to 1 produced an output of 0.055 and 0.29 respectively. Besides the output signal is also delayed by a phase equal to $m * 90° = 63°$

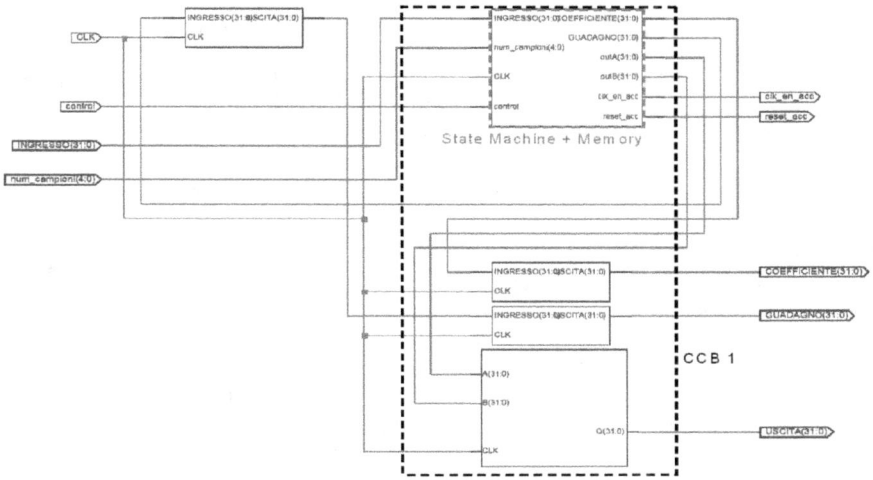

Fig. 4.3 The area CCB1 contains the State-Machine, the memory, and some blocks used to realize the computation of the coefficients in (4.33).

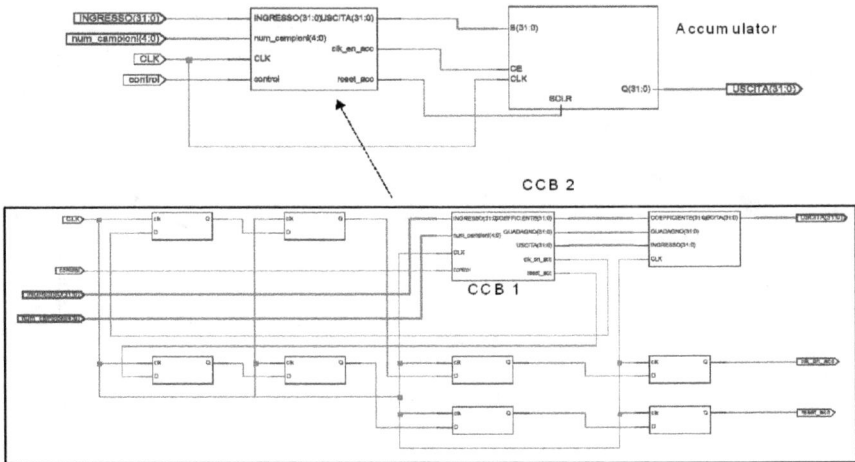

Fig. 4.4 Top hardware level that realizes the computation of the Fractional Differinte-gral Operator.

and $m * 90° = 27°$, equal to $35ms$ and $15ms$, respectively for the Integral Operator of fractional order 0.7 and 0.3, being the period of the input signal equal to $100ms = 200 * T_s = 200 * 0.5ms$, where 200 is the number of samples in a period and T_s is the sample time.

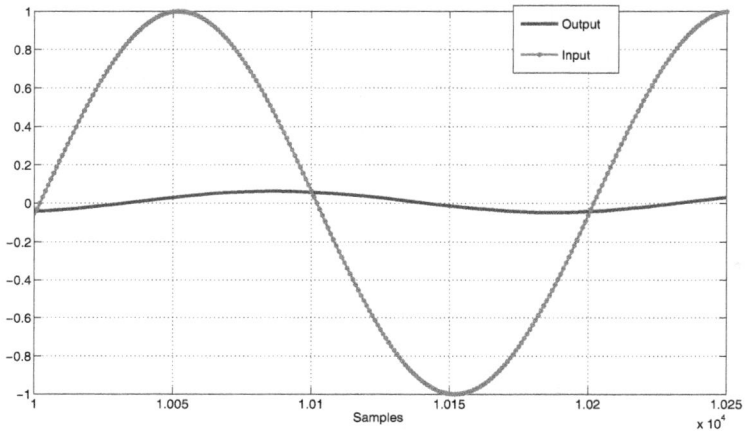

Fig. 4.5   Input and output data of a Fractional Integral Operator of order 0.7.

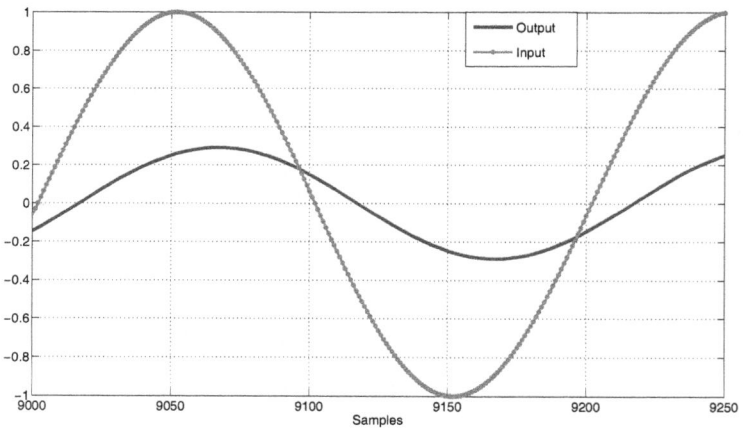

Fig. 4.6   Input and output data of a Fractional Integral Operator of order 0.3.

In Fig. 4.7 and 4.8 the bode diagrams of the two Integral Operator of fractional order 0.7 and 0.3, are shown respectively. In the work frequency range the fractional order behavioral is well approximated.

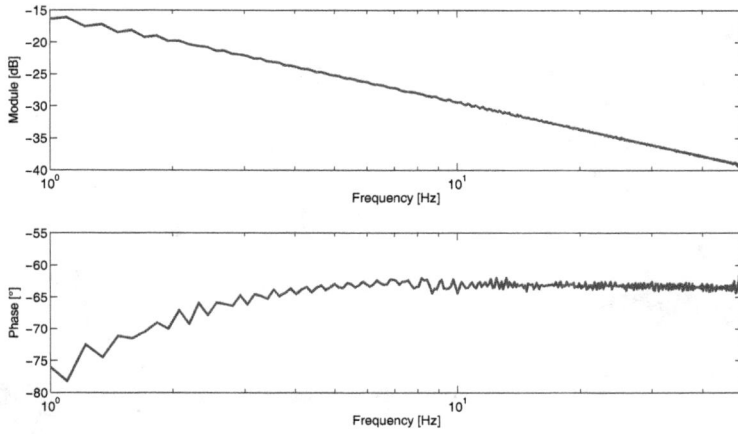

Fig. 4.7   Bode diagrams of a Fractional Integral Operator of order 0.7.

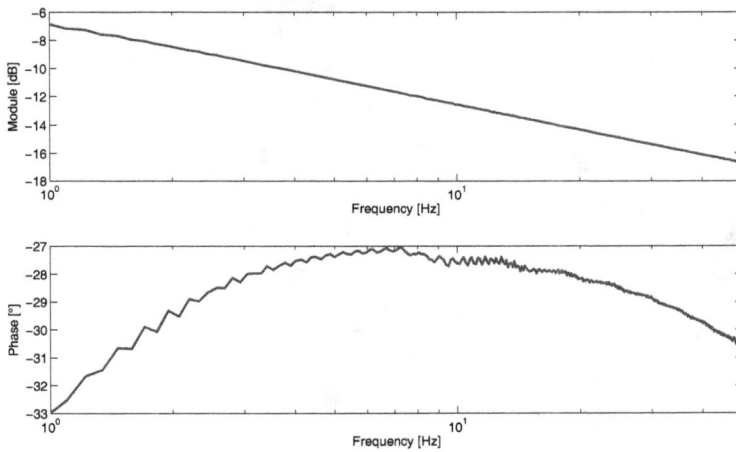

Fig. 4.8   Bode diagrams of a Fractional Integral Operator of order 0.3.

Chapter 5

Microprocessor Implementation and Applications

In this chapter the examples for digital implementation of fractional order controllers (FOC) will be given. A survey of several implementations and applications based on processor devices is presented. For illustration we present fractional order integrator, fractional order PD controller and fractional order ID controller.

5.1 Introduction

It is worth mentioning that, in general, the case of controller realization is not equivalent to the cases of simulation or numerical evaluation of the fractional integral and differential operators. In the case of controller implementation it is necessary to take into account some important considerations. First of all, the value of h, the step when dealing with numerical evaluation, is the value of the sample period T, and it is limited by the characteristics of the microprocessor-based system, used for the controller implementation, in two ways: (i) each microprocessor-based system has its own minimum value for the sample period, and (ii) it is necessary to perform all the computations required by the control law between two samples. Due to this last reason, it is very important to obtain good approximations with a minimal set of parameters. On the other hand, when the number of parameters in the approximation increases, it increases the amount of the required memory and speed too.

The key point in digital implementation of a FOC is the discretization of the fractional order operators. It is well known that, for interpolation or evaluation purposes, rational functions are sometimes superior to polynomials, roughly speaking, because of their ability to model functions with zeros and poles [Vinagre (2000a)].

The simplest and most straightforward method is the direct discretization using finite memory length expansion from Grünwald–Letnikov definition. In general, the discretization of fractional-order differentiator/integrator $s^{\pm r}$, $(r \in R)$ can be expressed by the generating function $s = \omega(z^{-1})$.

Using the generating function corresponding to the backward fractional difference rule, $\omega\left(z^{-1}\right) = \left(1 - z^{-1}\right)$, and performing the power series expansion (PSE) of $\left(1 - z^{-1}\right)^r$, the Grünwald–Letnikov formula for the fractional derivative of order r is obtained.

In any case, the resulting transfer function, approximating the fractional-order operators, can be obtained by applying the relationship:

$$Y(z) = T^{\mp r} \, \text{PSE}\left\{\left(1 - z^{-1}\right)^{\pm r}\right\} F(z), \qquad (5.1)$$

where T is the sample period, $Y(z)$ is the Z transform of the output sequence $y(nT)$, $F(z)$ is the Z transform of the input sequence $f(nT)$, and PSE$\{u\}$ denotes the expression, which results from the power series expansion of the function u.

Doing so gives:

$$D^{\pm r}(z) = \frac{Y(z)}{F(z)} = T^{\mp r} \, \text{PSE}\left\{\left(1 - z^{-1}\right)^{\pm r}\right\} \simeq T^{\mp r} P_p(z^{-1}), \qquad (5.2)$$

where $D^{\pm r}(z)$ denotes the discrete equivalent of the fractional-order operator, considered as processes.

By using the short memory principle [Podlubny (1999a)], the discrete equivalent of the fractional-order integro-differential operator, $(\omega(z^{-1}))^{\pm r}$, is given by

$$D^{\pm r}(z) = (\omega(z^{-1}))^{\pm r} = T^{\mp r} z^{-[L/T]} \sum_{j=0}^{[L/T]} (-1)^j \binom{\pm r}{j} z^{[L/T]-j}, \qquad (5.3)$$

where T is the sampling period, L is the memory length and $(-1)^j \binom{\pm r}{j}$ are binomial coefficients $c_j^{(r)}$, $(j = 0, 1, \dots)$ where [Dorčák (1994)]

$$c_0^{(r)} = 1, \qquad c_j^{(r)} = \left(1 - \frac{1 + (\pm r)}{j}\right) c_{j-1}^{(r)}. \qquad (5.4)$$

It is very important to note that the PSE scheme leads to approximations in the form of polynomials, that is, the discretized fractional order derivative is in the form of FIR filters. It should be mentioned that, at least for control purposes, it is not very important to have a closed-form formula for the coefficients, because they are usually pre-computed and stored in the memory of the microprocessor. In such a case, the most important is to have a limited number of coefficients because of the limited available memory of the microprocessor system.

Because of reasons menationed above, for directly discretizing s^r, ($0 < r < 1$), we shall concentrate on the IIR form of discretization where as a generating function we will adopt an Al-Alaoui idea on mixed scheme of Euler and Tustin operators [Al-Alaoui (1993, 1997)], but we will use a different ration between both operators. The mentioned new operator, raised to power $\pm r$, has the form [Petráš (2003b)]:

$$(\omega(z^{-1}))^{\pm r} = \left(\frac{1+a}{T} \frac{1-z^{-1}}{1+az^{-1}} \right)^{\pm r}, \qquad (5.5)$$

where a is ratio term and r is fractional order. The ratio term a is the amount of phase shift and this tuning knob is sufficient for most solved engineering problems, When $a = 0$ it is Euler rule, for $a = 1/7$ we get Al-Alaoui rule and for $a = 1$ we obtain well-known Tustin rule.

In expanding the above in rational functions, we will use the continued fraction expansion (CFE). It should be pointed out that, for control applications, the obtained approximate discrete-time rational transfer function should be stable and minimum phase. Furthermore, for a better fit to the continuous frequency response, it would be of high interest to obtain discrete approximations with poles an zeros interlaced along the line $z \in (-1, 1)$ of the z plane. The direct discretization approximations proposed in this section enjoy the desirable properties.

The result of such approximation for an irrational function, $\widehat{G}(z^{-1})$, can be expressed by $G(z^{-1})$ in the CFE form [Chen (2002)], [Vinagre (2003)]:

$$G(z^{-1}) \simeq a_0(z^{-1}) + \cfrac{b_1(z^{-1})}{a_1(z^{-1}) + \cfrac{b_2(z^{-1})}{a_2(z^{-1}) + \cfrac{b_3(z^{-1})}{a_3(z^{-1}) + \cdots}}}$$

$$= a_0(z^{-1}) + \frac{b_1(z^{-1})}{a_1(z^{-1})+} \frac{b_2(z^{-1})}{a_2(z^{-1})+} \cdots \frac{b_3(z^{-1})}{a_3(z^{-1})+} \cdots,$$

$$(5.6)$$

where a_i and b_i are either rational functions of the variable z^{-1} or constants. The application of the method yields a rational function, $G(z^{-1})$, which is an approximation of the irrational function $\widehat{G}(z^{-1})$.

The resulting discrete transfer function, approximating fractional-order operators $((\omega(z^{-1}))^{\pm r} \equiv D^{\pm r}(z^{-1}))$, can be expressed as [Petráš (2009b)]:

$$(\omega(z^{-1}))^{\pm r} \approx \left(\frac{1+a}{T}\right)^{\pm r} \mathrm{CFE}\left\{\left(\frac{1-z^{-1}}{1+az^{-1}}\right)^{\pm r}\right\}_{p,q}$$

$$= \left(\frac{1+a}{T}\right)^{\pm r} \frac{P_p(z^{-1})}{Q_q(z^{-1})}$$

$$= \left(\frac{1+a}{T}\right)^{\pm r} \frac{p_0 + p_1 z^{-1} + \cdots + p_m z^{-p}}{q_0 + q_1 z^{-1} + \cdots + q_n z^{-q}}, \qquad (5.7)$$

where T is the sample period, $\mathrm{CFE}\{u\}$ denotes the continued fraction expansion of u; p and q are the orders of the approximation and P and Q are polynomials of degrees p and q. Normally, we can set $p = q = n$.

Generally, the control algorithm can be based on canonical form of IIR filter, which can be expressed as follow

$$F(z^{-1}) = \frac{U(z^{-1})}{E(z^{-1})} = \frac{b_0 + b_1 z^{-1} + b_2 z^{-2} + \cdots + b_M z^{-M}}{a_0 + a_1 z^{-1} + a_2 z^{-2} + \cdots + a_N z^{-N}}, \qquad (5.8)$$

where $a_0 = 1$ for compatible with the definitions used in Matlab. Normally, we choose $M = N$.

Such control algorithm can be directly implemented to any processor based devices as for instance PLC or PIC. A direct form of such implementation using canonical form shown in Fig. 5.1 with input $e(k)$ and output

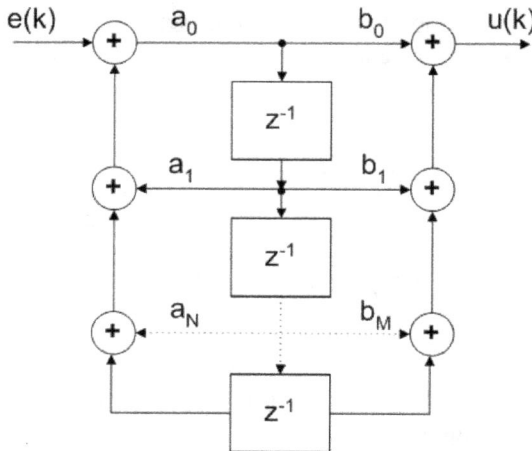

Fig. 5.1   Block diagram of the canonical representation of IIR filter form.

$u(k)$ range mapping to the interval $0 - U_{FOC}$ [V] is divided into two sections: initialization code and cyclic code. Pseudocode has the following syntax [Petráš (2005a)]:

```
(* initialization code *)
scale := 32752; % input and output
order := 5; % order of approximation (M=N)
U_FOC := 5; % input and output voltage range: 5[V], 10[V], ...
W:=1; % desired value of control loop
% Controller parameters:
a[0] := 1.0;  a[1] := ...;   a[2] := ...;  a[3] := ...;
a[4] := ...;  a[5] := ...;   b[0] := ...;  b[1] := ...;
b[2] := ...;  b[3] := ...;   b[4] := ...;  b[5] := ...;
% set variables to zero
loop i := 0 to order do
  s[i] := 0;
endloop

(* cyclic code *)
input_FOC := (REAL(input)/scale) * U_FOC;
in:=W-input_FOC;
feedback := 0; feedforward := 0;
loop i:=1 to order do
  feedback := feedback - a[i] * s[i];
  feedforward := feedforward + b[i] * s[i];
endloop
s[0] := in + a[0] * feedback;
out := b[0] * s[1] + feedforward;
loop i := order downto 1 do
  s[i] := s[i-1];
endloop
output_FOC:= INT(out*scale)/U_FOC;
% output limitations for actuator (saturation)
if output_FOC >= U_FOC then
   output_FOC = U_FOC;
if output_FOC <= 0 then
   output_FOC = 0;
```

The disadvantage with this solution is that the complete controller is calculated using floating point arithmetic.

The transfer function of the fractional order controller needs to be band-limited because the differentiation term leads to an infinity control effort value and great sensitivity to measurement noise. This impact can be limited by prefilter of first or second order connected to control loop. The goal for the prefilter is to correct the closed loop response of the feedback system because the actuator power capability is always limited. For some non band-limited inputs, the actuator becomes saturated and the saturation effect might cause undesirable non-linear phenomena.

The control algorithm is designed according to the control scheme in Fig. 5.2. The position algorithm uses discrete time steps k $(k = 1, 2, \dots)$ and consists of the following steps [Petráš (1999b)]:

(1) Reference value prefiltering:

For saturation effect limitation we used the first order digital prefilter with the constant correction factor, with the transfer function

$$H_p(z^{-1}) = \frac{W^*(z^{-1})}{W(z^{-1})} = \frac{k_f}{1 - k_f z^{-1}}. \qquad (5.9)$$

The transfer function of the prefilter (5.9) corresponds with the following difference equation

$$w^*(k) = w^*(k-1) + k_f(w(k) - w^*(k-1)), \qquad (5.10)$$

where $w(k)$ is the reference (desired) value, $w^*(k)$ is the prefiltered value and k_f is the prefilter constant.

(2) Control error computation:

The mathematical model of the difference term in control feedback loop has the form $E(z^{-1}) = W^*(z^{-1}) - Y(z^{-1})$ and the equivalent difference equation is

$$e(k) = w^*(k) - y(k), \qquad (5.11)$$

where $e(k)$ is control error value and $y(k)$ is the measured output value.

(3) Control effort value determination:

By applying the backward rule and PSE to the discrete transfer function of the fractional order controller we obtain a FIR form and the

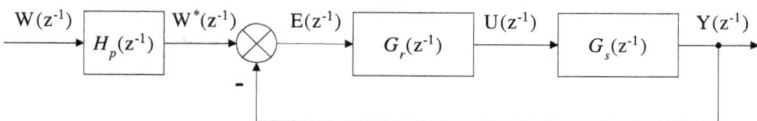

Fig. 5.2 Feed-back control loop with prefilter.

controller can be expressed as

$$C(z^{-1}) = K_p + \frac{T_i z^{[\frac{L}{T}]}}{T^{-\lambda} \sum\limits_{j=0}^{[L/T]} (-1)^j \binom{\lambda}{j} z^{[\frac{L}{T}]-j}} + \frac{T_d T^{-\delta} \sum\limits_{j=0}^{[L/T]} (-1)^j \binom{\delta}{j} z^{[\frac{L}{T}]-j}}{z^{[\frac{L}{T}]}}$$

(5.12)

From discrete transfer function (5.12), the control law is expressed:

$$u(k) = K_p e(k) + \frac{T_i}{T^{-\lambda} \sum\limits_{j=v}^{k} c_j^{(\lambda)} e(k-j)} + T_d T^{-\delta} \sum\limits_{j=v}^{k} c_j^{(\delta)} e(k-j),$$

(5.13)

where binomial coefficients $c_j^{(\delta)}$ and $c_j^{(\lambda)}$ are calculated from the recurrent equation (5.4) and v is defined in context of the relation (5.27) for improving the effectiveness of the numerical algorithm by applying the short memory principle [Podlubny (1999a)]. On the other hand, besides the sample period T, the control quality is also influenced by the length of the "short memory".

In the case of IIR form of the fractional order controller, the control law can be expressed as:

$$u(k) = K_p e(k) + T_i T^{\lambda} \frac{\sum_{j=0}^{r} c_j e(k-j)}{\sum_{i=0}^{r} d_j e(k-i)} + \frac{T_d}{T^{\delta}} \frac{\sum_{j=0}^{p} a_j e(k-j)}{\sum_{i=0}^{p} b_i e(k-i)},$$

(5.14)

being p, q, r, s and a_j, b_i, c_j, d_i the orders and coefficients of the polynomials, respectively.

Equation (5.14) can be rearranged and expressed as one difference equation with $u(k)$ on the left side and finite difference on the right side. This approach does not need the short memory principle. We need only few samples from history for computation a new one.

Some other discrete algorithms for the digital realization of the fractional-order controller can be used, such as for example the second order of the prefilter or the CFE method of Tustin rule or Simpson rule (various value of a in relation (5.7)) for approximation of the discrete controller transfer function in the form of the IIR filter.

5.2   Fractional Controller Realized by PIC Processor

The proposed realization of the fractional controller is based on the microprocessor PIC16F876. The block diagram of the mentioned realization is depicted in Fig. 5.3. Setting of the controller parameters is realized via serial port RS232 by personal computer (PC). Basic features of processor PIC16F876 [Microchip (2009)]: 256 data memory bytes, and 368 bytes of user RAM, an integrated 5-channel 10-bit AD converter. Two timers. Precision timing interfaces are done through two CCP modules and two PWM modules. PIC16F876 has 22 I/O pins. The operating speed is 20 MHz and power supply voltage is 5 V (fairly well filtered).

Basic features of realized digital FOC: The range of input signal (control error) is 0 - 5 V and can be adjusted for any sensor output. The output signal range (control value) is 0 - 5 V. The reference voltage range for required value setting is also 0 - 5 V. The controller was designed for polynomial degree $p = q = 3$ and for sample period T (0.001 - 120) sec. This controller was build as a prototype for testing and measurements (see [Petráš (2003c)].)

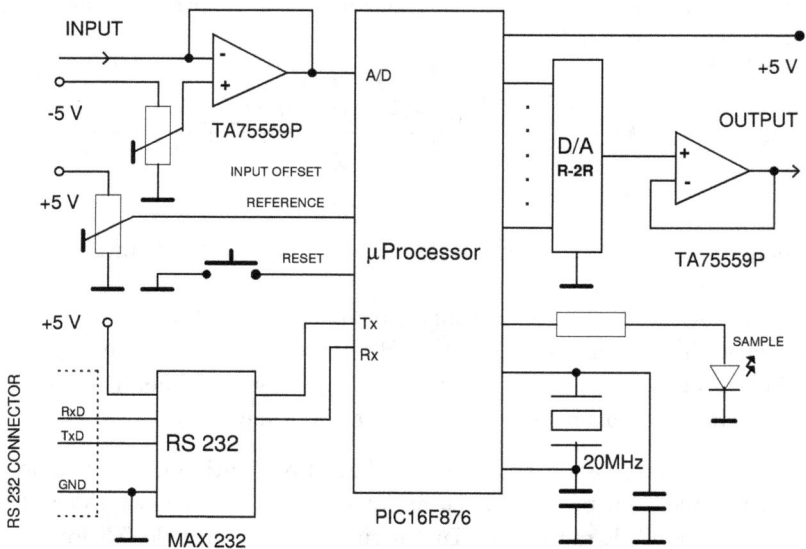

Fig. 5.3   Block diagram of digital FOC based on PIC.

5.2.1 *Fractional-Order Integrator*

This section uses the practical case described in [Petráš (2002a)]. We realized the fractional half-order integrator as a particular case of fractional order PID controller, which has continuous transfer function:

$$C(s) = \frac{1.4374}{s^{0.5}}.$$

The resulting transfer function of half-order integrator was obtained by (5.7) and for $T = 0.001$ sec and $a = 1/7$ has the following form:

$$C(z) = 1.4374 \frac{49z^3 - 49z^2 + 7z + 1}{1657z^3 - 2603z^2 + 1048z - 63}. \tag{5.15}$$

5.2.2 *Measured Results*

The transfer function (5.15) was rewritten to difference equation and was coded by PIC Basic and then loaded to PIC memory [Petráš (2003c)]. As the testing signals we used a square impulses (unit - step response) and sinus. Both signals had amplitude 1V and frequency 100 Hz. The following figures Fig. 5.4 and Fig. 5.5 present the measured results (by digital oscilloscope). These results are comparable with results obtained by analog FOC in [Petráš (2002a)].

5.3 Temperature Control of a Solid by PC and PCL 812

5.3.1 *Model of Controlled System*

The mathematical model used for the system to be controlled is a two-term differential equation of the fractional-order of the form:

$$b_1 D_t^\beta y(t) + b_0 y(t) = u(t), \tag{5.16}$$

for which the parameters b_1, b_0 and β were obtained by an identification method based on the measured step response of the system (see Fig. 5.6) and the minimization of the quadratics criteria

$$J = \frac{1}{M+1} \sum_{i=0}^{M} |y_i^* - y_i|^2 \tag{5.17}$$

being y_i^* the measured values, y_i the model values and $M+1$ the number of measurements.

In this case, the obtained parameters are [Petráš (1998)]:

$$b_1 = 39.69; \qquad b_0 = 0.598; \qquad \beta = 1.25$$

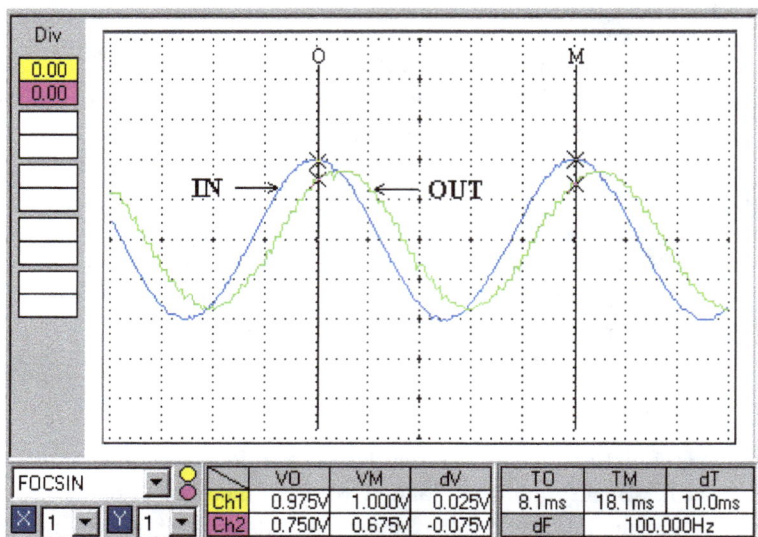

Fig. 5.4   Time response to sin function.

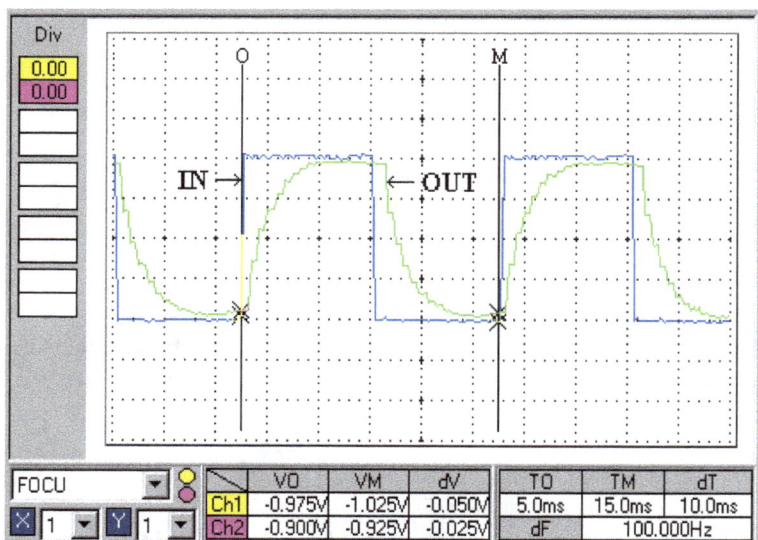

Fig. 5.5   Time response to unit-step function.

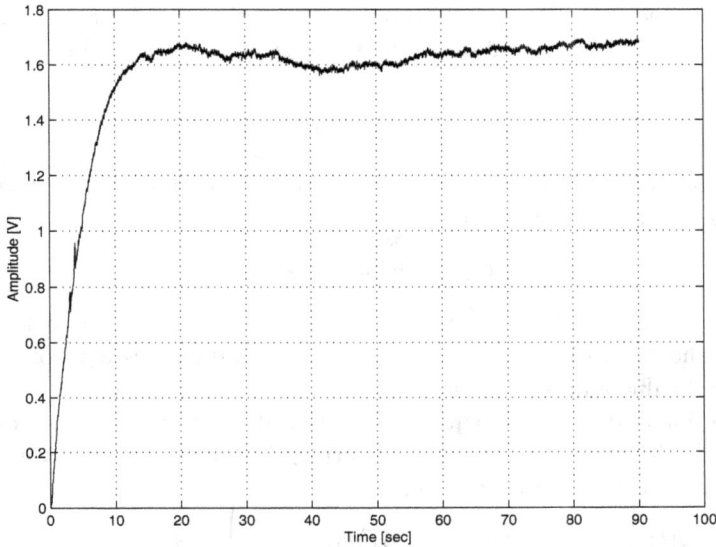

Fig. 5.6 Unit-step response of controlled object.

So, the continuous transfer function used for controller design is

$$G(s) = \frac{1}{39.69s^{1.25} + 0.598}. \tag{5.18}$$

This mathematical model was used in [Petráš (1998)] for fractional controller design, and an alternative integer order model was used for traditional PD controller design with comparison purposes.

The alternative integer order model has the form of first order system represented by the following transfer function:

$$\widehat{G}(s) = \frac{1}{20.14s + 0.598}. \tag{5.19}$$

This integer order system was used for comparison of control performance between classical PD controller and fractional PD^δ controller with transfer function [Dorčák (1994)], [Petráš (1998)]:

$$C(s) = K + T_d s^\delta, \tag{5.20}$$

where K, T_d and δ are controller parameters.

5.3.2   *Controller Parameters Design and Implementation*

The controller design was done in [Petráš (1998)], according to the method (poles placement [Dorf (1990)]) described in [Petráš (1999b)], for obtaining a stability measure $St \approx 2.0$. The obtained fractional-order PD^{δ} controller designed for the fractional-order model (5.18) has the continuous transfer function:

$$C(s) = 64.47 + 48.99 s^{0.5}. \tag{5.21}$$

Parameters of integer order PD controller was designed by the same method and controller has the following transfer function:

$$\widehat{C}(s) = 64.47 + 12.39 s. \tag{5.22}$$

Let us consider the single input - single output (SISO) feedback control system shown in Fig. 5.7, where W is required value, E is control error, U is control value and Y is actual value.

Fractional differential equation of closed control loop, depicted in Fig. 5.7, with fractional model of controlled system and fractional PD^{δ} controller, has the following form:

$$a_1\,{}_0D_t^{\beta}y(t) + T_d\,{}_0D_t^{\delta}y(t) + (a_0 + K)y(t) = K\ w(t) + T_d\,{}_0D_t^{\delta}w(t). \tag{5.23}$$

The comparison of the control quality was done by applying both controllers to the real object in [Petráš (2000)], where for the control quality arbitration were chosen the following criteria: control surface; overshoot; settling time, which were measured the values of these criteria for total time 2 min for unit change of required value from 0 to 2 V. It correspondences to temperature change about $100^{o}C$. The obtained results confirmed the results from previous works that fractional order controller is more eligible for control of the real object. In the next part we will consider only fractional order controller.

Since the advantages of using a fractional controller in this particular case were shown in [Petráš (1998)], in this section we compare two possible realizations of the fractional order PD^{δ} controllers.

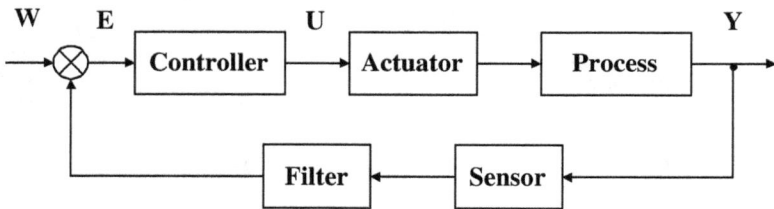

Fig. 5.7   General SISO feedback loop system.

By applying the backward rule and power series expansion (Grünwald–Letnikov formula) the discrete transfer function of the controller can be expressed of the form:

$$C_R(z) = K + T_d \frac{T^{-\delta} \sum_{j=0}^{L/T} (-1)^j \binom{\delta}{j} z^{\frac{L}{T}-j}}{z^{\frac{L}{T}}}, \tag{5.24}$$

being T the sample period, L the memory length (see [Podlubny (1999a)]) and $L/T + 1$ the number of coefficients.

By applying the Tustin trapezoidal rule and the continued fraction expansion the discrete transfer function of the controller can be expressed of the form:

$$C_T(z) = K + T_d \left(\frac{2}{T}\right)^\delta \frac{P_p(z^{-1})}{Q_q(z^{-1})}. \tag{5.25}$$

For implementing the controllers a position algorithm with reference digital prefiltering has been used. This algorithm consists of the several steps (calculating of control error, calculating the control value, etc.). In this section we present only control laws, which are described as [Petráš (2002b)]:

a) In the case of $C_R(z)$ the control law can be expressed as:

$$u(k) = Ke(k) + \frac{T_d}{T^\delta} \sum_{j=n}^{k} c_j e(k - j), \tag{5.26}$$

being c_j the binomial coefficients and n, by applying the short memory principle [Podlubny (1999a)], defined by:

$$\begin{aligned} n = 0; & \quad k < L/T \\ n = k - L/T; & \quad k > L/T \end{aligned} \tag{5.27}$$

b) In the case of $C_T(z)$ the control law can be expressed as:

$$u(k) = -\frac{1}{b_0} \sum_{l=1}^{q} b_l u(k - l)$$

$$+ \frac{1}{b_0} \left[K \sum_{l=0}^{q} b_l e(k - l) + \frac{T_d}{T^\delta} \sum_{m=0}^{p} a_m e(k - m) \right], \tag{5.28}$$

being p, q and a_m, b_l the orders and coefficients of the polynomials $P_p(z^{-1})$ and $Q_q(z^{-1})$, respectively.

5.3.3   *Experimental Setup and Results*

The equipment for experiments is shown in Fig. 5.8. The system to be controlled is a heat solid (electric radiator). Temperature is measured by a radiating pyrometer, filtered by an analogue active filter, and driven to host PC by a PCL 812 card. Control signal from analogue output on the PCL card is connected to the actuator (thyristor changer) where 0-5V signal is changed to 20-220V. Reference value follows the law:

$$w(^o) = \frac{330}{5}w(V) + 20 \tag{5.29}$$

The estimated transfer function of the analogue filter is

$$F(s) = \frac{22}{s + 40}, \tag{5.30}$$

which has been used for filtering the noisy signal from pyrometer.

In experiments the following parameters have been used:

- $T = 1$ sec, ($\simeq 1\%$ of the system rise time);
- $L = 100$ (order of the FIR filter);
- $k_f = 0.5$; (prefilter parameter);
- $p = q = 4$ (order of the IIR filter);

With these parameters the implemented controllers are [Vinagre (2001)]:

$$C_R(z) = 64.47 + 48.99 \frac{\sum_{k=0}^{100} (-1)^k \binom{0.5}{k} z^{100-k}}{z^{100}},$$

$$C_T(z) = 64.47 + 48.99 \frac{0.316z^4 - 1.038z^3 + 1.248z^2 - 0.645z + 0.119}{0.256z^4 - 0.639z^3 + 0.488z^2 - 0.078z - 0.027}.$$

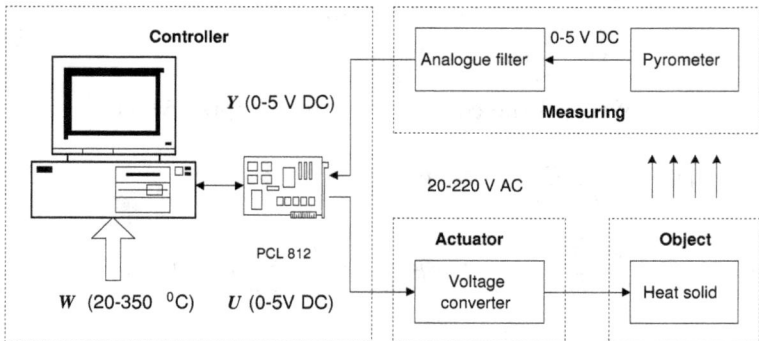

Fig. 5.8   Block diagram of experimental setup.

The transfer function of the digital prefilter is:

$$H_p(z) = \frac{0.5}{1 - 0.5z^{-1}}.$$

This prefiltering can improve control loop performances e.g. less over-shoot, etc. Usually, it is suitable use the first order system as a prefilter with time constant which corresponds to the time constant of controlled systems.

The simulation results are obtained by applying the controllers $C_R(z)$ and $C_T(z)$ to the process transfer function. Presented results consider the ideal case of Fig. 5.8, that is: no actuator saturation and unity feedback. Simulated step responses of the controlled system with controllers $C_R(z)$ and $C_T(z)$ are shown in Fig. 5.9. In this figure it can be observed that the performances for both controllers are identical.

Measured step responses of the controlled system with controllers $C_R(z)$ and $C_T(z)$ are shown in Fig. 5.10. As in the case of simulations, the almost identical performances obtained with both controllers can be observed.

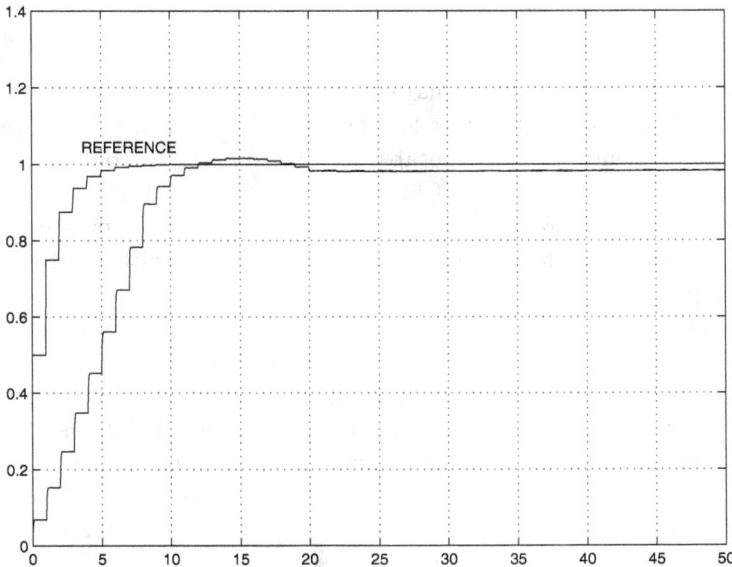

Fig. 5.9 Simulated unit step response (x: Time [sec]; y: Amplitude [V]).

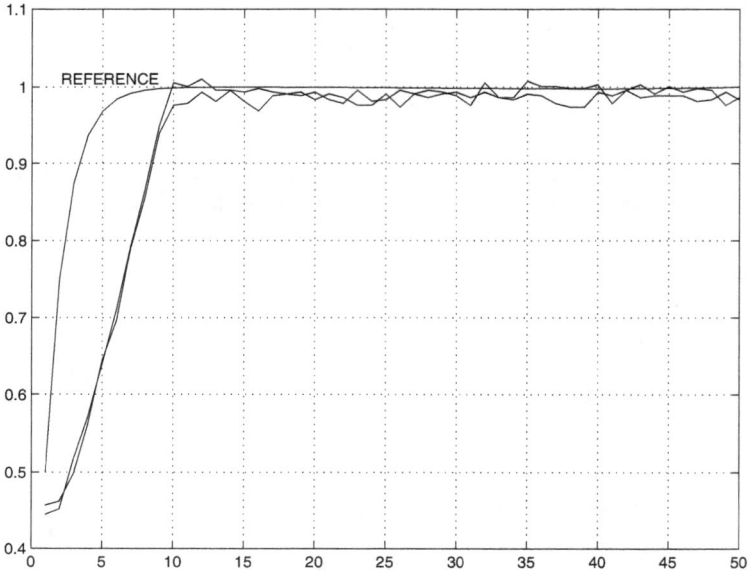

Fig. 5.10   Measured unit step response (x: Time [sec]; y: Amplitude [V]).

Control algorithm used in this experiment was programmed in computer language Pascal 6.0 with driver for the PCL 812 card. The disadvantage of this implementation was the absence of software environment at that time. All necessary routines af for example driver for PLC card, timing, control loop, etc. had to be programmed and coded. At present time it can be solved easily, for instance in Matlab/Simulink with the Real-Time Workshop toolbox.

The advantage of using the second method for implementation is clear: while the controller $C_R(z)$ is a FIR filter of order 100, the controller $C_T(z)$ is an IIR filter of order 4. From the obtained results it can be concluded that for implementing the digital fractional controller is highly interesting to use Tustin rule and continued fraction expansion, because it reduces, without performance degradation, the digital system requirements. This means that the implementation of $C_T(z)$ has reduced requirements in memory and computation time.

5.4 Temperature Control of a Heater by PLC BR 2005

5.4.1 *Model of Controlled System*

In this section will be described the laboratory object - electrical heater and control setup of whole system. Electrical heater consists of electrical spiral and fan. In front of the heater is temperature sensor and this heater is connected to power actuator.

The mathematical model used for the system to be controlled can be expressed by the following transfer function:

$$G(s) = \frac{k}{\tau s + 1}. \tag{5.31}$$

Parameters of the model (5.31) could be obtained by graphical identification method [Dorf (1990)] and have the following values (see Fig. 5.11):

$$\tau = 75, \quad k = 0.57. \tag{5.32}$$

For real measurement was used sampling period $T = 5$ sec. In Fig. 5.12 is depicted unit step response of the heater recalculated to temperature according to transfer characteristic of the sensor. The initial condition was $y(0) = 26^{\circ}C$ (temperature in room).

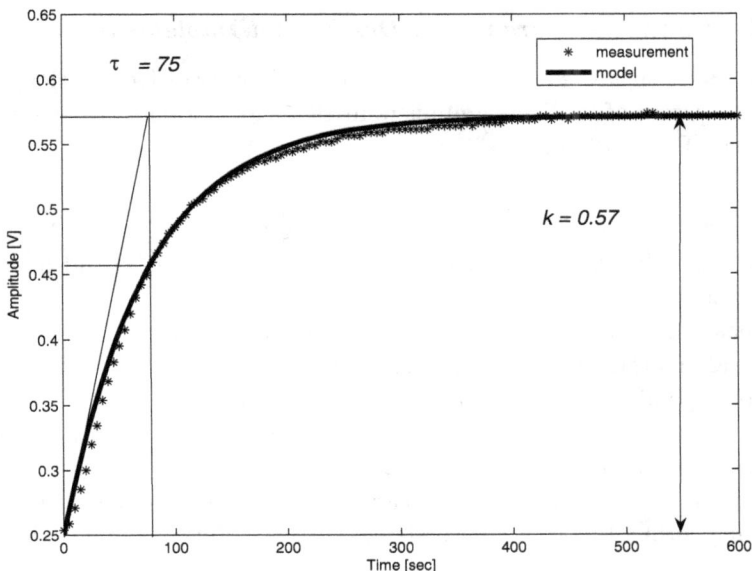

Fig. 5.11   Unit step responses (measured and model).

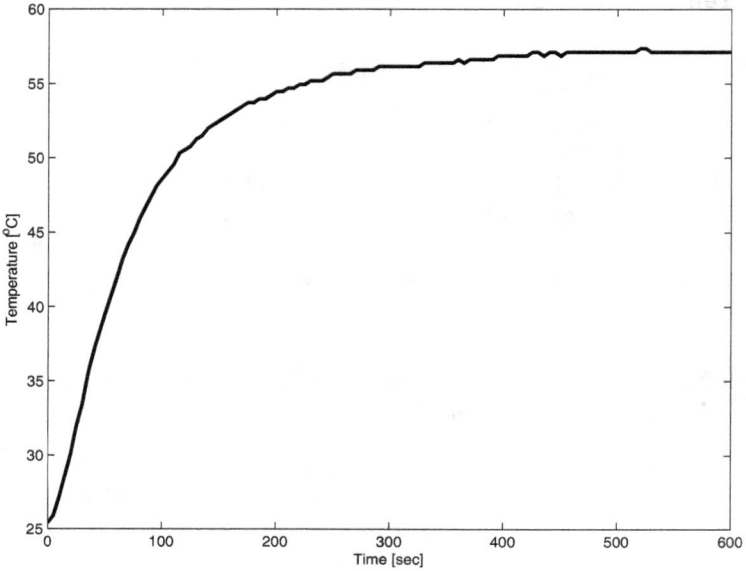

Fig. 5.12   Unit step responses (measured).

5.4.2   *Controller Parameters Design and Implementation*

H. W. Bode suggested an ideal shape of the loop transfer function in his work on design of feedback amplifiers in 1945. Ideal loop transfer function has form [Bode (1949)]:

$$L(s) = \left(\frac{s}{\omega_{gc}} \right)^{\alpha}, \tag{5.33}$$

where ω_{gc} is desired crossover frequency and α is slope of the ideal cut-off characteristic.

Phase margin is $\Phi_m = \pi(1 + \alpha/2)$ for all values of the gain. The amplitude margin A_m is infinity. The constant phase margin 60°, 45° and 30° correspond to the slopes $\alpha = -1.33$, -1.5 and -1.66.

The Nyquist curve for ideal Bode transfer function is simply a straight line through the origin with $\arg(L(j\omega)) = \alpha\pi/2$.

Bode's transfer function (5.33) can be used as a reference system in the following form [Astrom (2000); Manabe (1961); Petráš (2002a); Vinagre (2004)]:

$$G_c(s) = \frac{A}{s^{\alpha} + A}, \qquad G_o(s) = \frac{A}{s^{\alpha}}, \qquad (0 < \alpha < 2), \tag{5.34}$$

where $G_c(s)$ is transfer function of closed loop and $G_o(s)$ is transfer function in open loop.

General characteristics of Bode's ideal transfer function are:

(a) Open loop:

- Magnitude: constant slope of $-\alpha 20 dB/dec$;
- Crossover frequency: a function of A;
- Phase: horizontal line of $-\alpha \frac{\pi}{2}$;
- Nyquist: straight line at argument $-\alpha \frac{\pi}{2}$.

(b) Closed loop:

- Gain margin: $A_m = \infty$;
- Phase margin: constant: $\Phi_m = \pi \left(1 - \frac{\alpha}{2}\right)$;
- Step response: $y(t) = At^\alpha E_{\alpha,\alpha+1}\left(-At^\alpha\right)$,
  where $E_{a,b}(z)$ is the Mittag-Leffler function of two parameters [Podlubny (1999a)].

For the FOC design we will use an idea which was proposed by Bode [Bode (1949)] and for first time used to the motion control described by Tustin [Tustin (1958)]. This principle was also used by Manabe to induction motor speed control [Manabe (2002)].

We will design the controller, which give us a step response of feedback control loop with overshoot independent of payload changes (iso-damping). In the frequency domain point of view it means phase margin independent of the payload changes.

Phase margin of controlled system is [Monje (2008); Vinagre (2000)]:

$$\Phi_m = \arg\left[C(j\omega_g)G(j\omega_g)\right] + \pi, \tag{5.35}$$

where $j\omega_g$ is the crossover frequency. Independent phase margin means in other words constant phase. This can be accomplished by controller of the form:

$$C(s) = k_1 \frac{k_2 s + 1}{s^\mu}, \quad k_1 = 1/k, \quad k_2 = \tau. \tag{5.36}$$

Such controller gives a constant phase margin and obtained phase margin is

$$\Phi_m = \arg\left[C(j\omega)G(j\omega)\right] + \pi = \arg\left[\frac{k_1 k}{(j\omega)^{(1+\mu)}}\right] + \pi$$

$$= \arg\left[(j\omega)^{-(1+\mu)}\right] + \pi = \pi - (1 + \mu)\frac{\pi}{2}. \tag{5.37}$$

For our parameters of controlled object (5.32) and desired phase margin $\Phi_m = 45^o$, we get the following constants of the fractional order controllers

(5.36): $k_1 = 1.75$, $k_2 = 75$ and $\mu = 1.5$. With these constants we obtain a modified fractional $I^\lambda D^\delta$ controller, which is a particular case of the $PI^\lambda D^\delta$ controller and has the form:

$$
\begin{aligned}
C(s) &= \frac{\tau}{k} s^{-0.5} + \frac{1}{ks^{1.5}} \\
&= \frac{1}{s}\left(\frac{\tau}{k} s^{0.5} + \frac{1}{ks^{0.5}}\right) \\
&= \frac{131.57\sqrt{s}}{s} + \frac{1.75}{s\sqrt{s}} \\
&= \frac{1}{s}(K_d s^\delta + K_i s^{-\lambda}),
\end{aligned}
\tag{5.38}
$$

where K_d, K_i, δ, and λ are the fractional $I^\lambda D^\delta$ controller parameters.

According to relation (5.37), by using a controller (5.38), we can obtain a phase margin:

$$
\Phi_m = \arg\left[C(j\omega)G(j\omega)\right] + \pi = \pi - (1.5)\frac{\pi}{2} = 45^o,
$$

which was desired phase margin specification.

5.4.3   Experimental Setup and Results

Experimental setup used for identification of the object and its control is depicted in Fig. 5.13.

Temperature was measured by semiconductor sensor LM35 with linear characteristic, sensitivity $10mV/^oC$ and range from -35^oC to 150^oC. As an actuator was used a thyristor converter controlled by PIC processor with input voltage range $0 - 5V$ DC and output voltage range $0 - 230V$ AC.

For implementation of the fractional order controller (5.38) on PLC we can use approximation based on the CFE in form of IIR filter and algorithm described in first section of this chapter. For the sampling period $T = 1$ sec and by using the Tustin rule as a generation function, we get the discrete approximation of the fractional order controller (5.38) in the following form:

$$
C(z^{-1}) = \frac{\begin{array}{c}93.6664 + 48.0756z^{-1} - 149.0885z^{-2} - 63.8758z^{-3} \\ +59.5138z^{-4} + 17.0560z^{-5} - 3.5300z^{-6}\end{array}}{\begin{array}{c}1.0000 - 0.5001z^{-1} - 1.6111z^{-2} + 0.6806z^{-3} \\ +0.6459z^{-4} - 0.1841z^{-5} - 0.0382z^{-6}\end{array}}.
\tag{5.39}
$$

Above fractional order controller was implemented in PLC BR 2005 as a task FOC in first cyclic class with sampling time $T = 1$ sec. To input analogue modul was connected the semiconductor sensor LM35 and to output

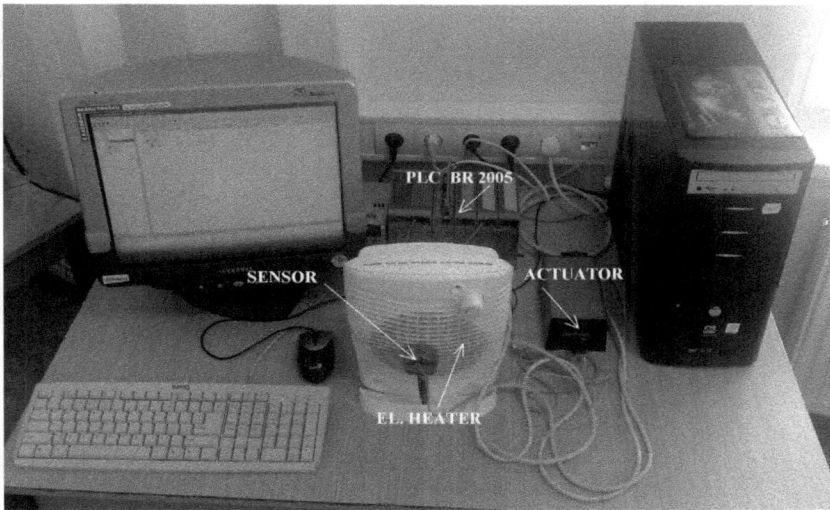

Fig. 5.13   Experimental setup in laboratory.

analogue modul was connected the actuator. The PLC was controlled via PC through RS 232. The code was programmed in software Automation studio in language Automation Basic and updated to memory of the PLC.

As we can see in Fig. 5.14 the Bode plots of open control loop with the model (5.31) with parameters (5.32) and controller (5.39) have phase margin $\Phi_m \approx 44.28$, phase 135° and magnitude slope -30dB/dec.

Simulation of the closed control loop with the model (5.31) with parameters (5.32) and controller (5.39) shows (see Fig. 5.15) that unit-step response has almost 80 % overshoot and we should consider a prefilter for desired value to eliminate this overshoot.

Suggested prefilter transfer function has the form:

$$H_p(s) = \frac{1}{38s + 1}. \tag{5.40}$$

Discrete equivalent of the prefilter (5.40) obtained through c2d() for sampling period $T = 1$ sec is

$$H_p(z) = \frac{0.026}{z - 0.974}. \tag{5.41}$$

As we can see in Fig. 5.16, the prefilter (5.41) helped us eliminate overshoot but it is important to note that the settling time changed from 25 sec to almost 150 sec. Sometimes it is not acceptable for such kind of process but usually overshoot is superior for heat processes then settling time. Overshoot of temperature value may destroy the object.

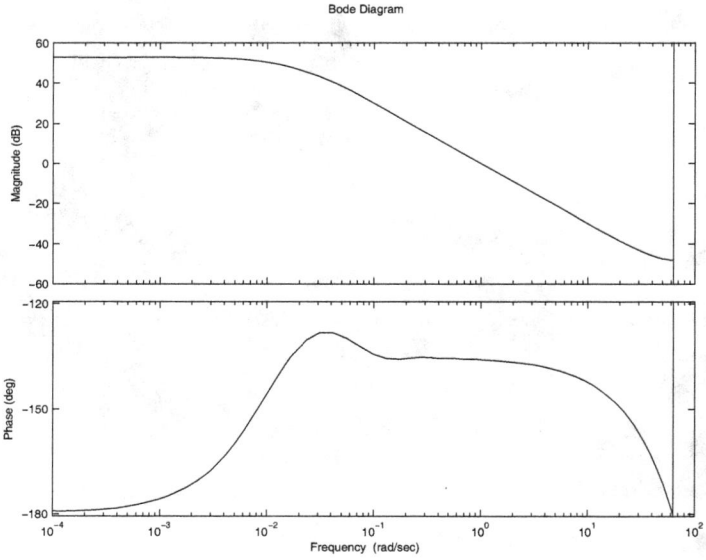

Fig. 5.14   Bode plots (simulated).

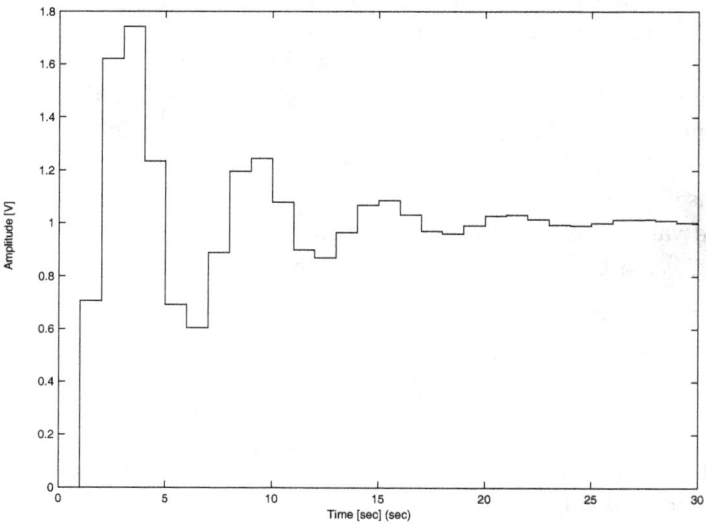

Fig. 5.15   Unit step responses (simulated).

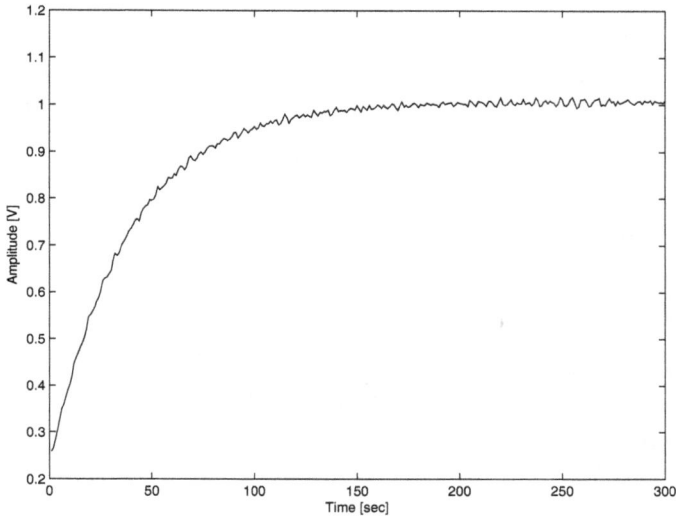

Fig. 5.16 Unit step responses with prefilter (measured).

5.5 Concluding Remarks

As it has been shown in this chapter, implementation of the fractional order controller on processor based devices is possible but there are many important things which should be solved before its application to practice.

In case of PIC processor could be a problem with arithmetic operations in floating point numbers and the speed of mathematical calculation as well. This can be solved with coprocessor. Using the PC and PCL card are suitable mostly for laboratory experiments.

Some disadvantage we can mention also in case of PLC implementation, as for example limitation to memory and processor speed (depends on type of PLC). On the other hand, example in this chapter has shown us, how we can implement the fractional order controller in PLCs, which are widely used in industry at present time.

Chapter 6

Field Programmable Analog Array Implementation

The aim of this chapter is to propose an analog implementation of $PI^\lambda D^\mu$ controller based on Field Programmable Analog Arrays (FPAAs). The proposed approach guarantees a good degree of approximation and the possibility to easily fix the integrative and derivative order. The frequency analysis of the reported examples shows the feasibility and reliability of the proposed approach.

6.1 The FPAAs Development System

Digital circuits are usually used when high-accuracy, high-complexity signal processing algorithms have to be implemented. This approach is typically applied to low-frequency signals when power consumption is not critical.

When, on the contrary, high-frequency signals are involved and low power dissipation is required, the analog approach is more appealing. The drawback of the analog approach is related to the accuracy of the circuits and their reprogrammability.

The FPAA technology, as for its digital counterpart FPGA, provides a basic instrument for the development of dynamically reconfigurable analog circuits. Using this kind of device it is possible to reprogram the entire circuit dynamics, keeping the structure fixed but changing the parameters [Caponetto (2007b)]. The reprogrammability features of FPAA can also be used to adapt the circuit to changing external conditions due to noise or changes in the operating conditions of the system being controlled. The circuit configurations can be changed, where components such as operational amplifiers, capacitors, resistors, transconductors, and current mirrors can easily be fixed and connected, both at a low and at a high level. In the latter case, user-friendly tools, for example to design audio amplifiers, are

115

available in order to reduce the time to market products. In a way, FPAA is a new version of an analog computer, maintaining the same target of providing analog circuits with flexibility. Therefore, the two main characteristics of FPAA are the possibility to translate complex analog circuits into a set of low-level functions and the capability to place analog circuits under real-time software control within the system. For these reasons FPAA is used here to implement non integer order circuits with programmable features.

The device used in this paper is the FPAA AN221E04, produced by Anadigm, intergraded on the development board AN221K04, produced by the same company. The software development tool is the AnadigmDesigner2 (Anadigm). The core of the AN221E04 is a two by two matrix of block named CAB (Configurable Analog Block) that can be connected among them and with external I/O blocks. In Fig. 6.1 the scheme of the Anadigm device is reported.

Each CAB contains a digital comparator, an analog comparator, two operational amplifiers and a series of capacitors. The FPAA technology is, in fact, mainly based on switched capacitor technology. The CAB blocks are surrounded by the other elements of the device. One section is dedicated to clock management, another to I/O signals and a digital section is devoted to the IC configuration and dynamic reprogrammability. The digital section is based on a look-up table mapping the interconnections inside the IC. The look-up-table allows the dynamic reprogramming of the FPAA. In fact it is possible to connect the AN221E0 with an external micro-controller and to change the values of the look-up-table on the fly. These values will be applied at the next clock cycle. Other features characterize the AN221E0 but are not described here because we did not use them in our experiments.

The software development tool makes it possible to connect the FPAA with the I/O ports and to design the desired circuit, by using pre-defined blocks. There are a lot of blocks, named CAM (Configurable Analog Module) that differ in functionality and therefore in resource requirement. In the following only those blocks that will be utilized in our experiment are described.

The first one, and the simplest one, is the GAININV block that permits us to fix a gain, (in the range $0.01 \div 100$), and an internal clock for switch driving.

A further block is the SUMINV. This CAM creates a full cycle inverting the summing stage with up to three inputs. Each input branch has a programmable gain. This CAM has continuous input and continuous output

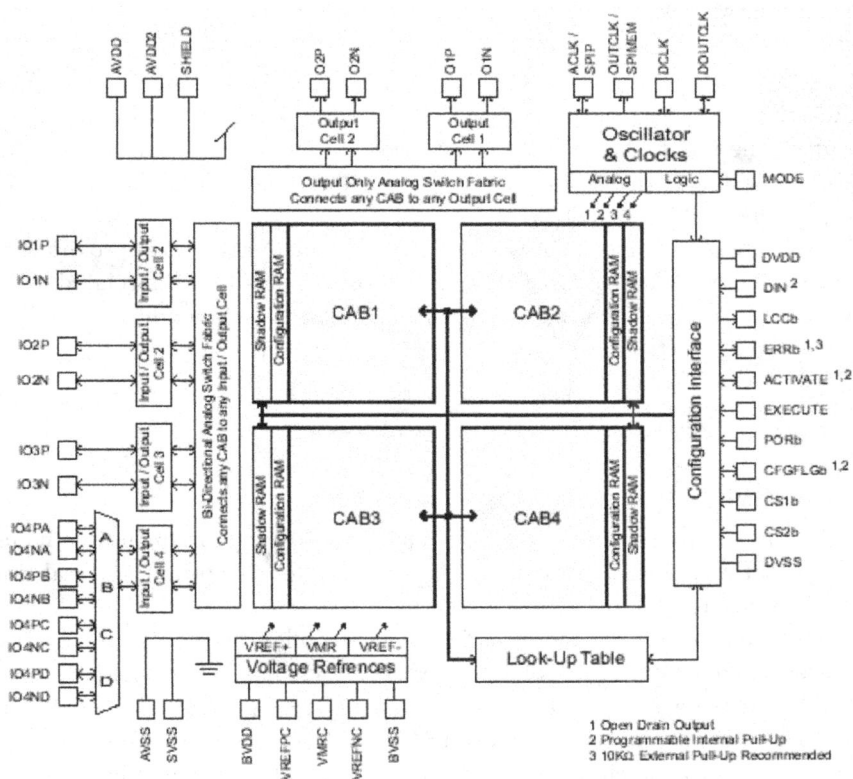

Fig. 6.1 Block scheme of the device AN221E04.

that are always valid. The transfer function for this CAM is:

$$V_{Out}(s) = -G_1 V_{Input1}(s) - G_2 V_{Input2}(s) - G_3 V_{Input3}(s) \qquad (6.1)$$

The numbered G_i variables are the gains of the various input branches and the numbered V_{Input} variables are the input voltages at the various input branches.

Another CAM used is the DIFFERENTIATOR. This CAM creates an inverting differentiator with a programmable differentiation constant. This CAM operates as a circuit that generates an output based on the change of input during one phase and holds that output through the next phase or a circuit that generates an output based on the change of input during every phase. The transfer function for this CAM is:

$$V_{Out}(s) = -K \times \frac{\Delta V_{in}}{\Delta t} \qquad (6.2)$$

K is the differentiation constant and ΔV_{in} is the change in the input voltage during the input phase of the clock when Δt is the length of one half the clock period. If the input voltage does not change before the end of the input phase, ΔV_{in} equals zero.

The most important CAM used to realize the Oustaloup approximation is the BILINEAR FILTER CAM. The choice of this CAM is related to its capability to implement zero-pole filters. Its transfer function is:

$$\frac{V_{out}(s)}{V_{in}(s)} = -\frac{G_{HF}(s + 2\pi f_z)}{(s + 2\pi f_p)} \qquad (6.3)$$

with the condition:

$$G_{DC} = \frac{f_z}{f_p} G_{HF} \qquad (6.4)$$

where f_z is zero frequency, f_p pole frequency, G_{HF} high frequency gain and G_{DC} static gain. The described transfer function is optimal for realizing the basic cell of Oustaloup interpolation.

In order to fix poles and zeros frequencies configuration a Matlab routine based on Oustaloup interpolation (see eqs. 1.45) has been developed.

The routine starting point is the constraint $G_{DC} = \frac{f_z}{f_p} G_{HF}$. This formula leaves one parameter between G_{DC} and G_{HF} unconstrained.

The equivalent circuit of the bilinear filter is shown in Fig. 6.2. All the bilinear filter CAM parameters, essentially the values of the capacitors, have been fixed taking into account the Oustaloup approximation via the formulas:

$$f_p = \frac{f_c}{\pi} \frac{C_3}{(2C_4 + C_3)} \qquad (6.5)$$

$$f_z = \frac{f_c}{\pi} \frac{C_1}{(2C_2 + C_1)} \qquad (6.6)$$

$$G_{DC} = \frac{C_1}{C_3} \qquad (6.7)$$

$$G_{HF} = \frac{(2C_2 + C_1)}{(2C_4 + C_3)} \qquad (6.8)$$

The entire $PI^\lambda D^\mu$ controller CAM scheme, obtained via the Anadigm development tool, is reported in Fig. 6.3.

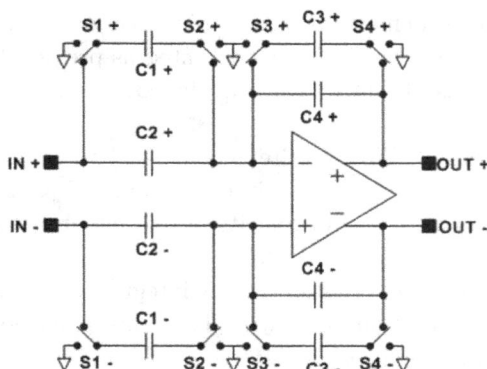

Fig. 6.2   Scheme of the pole-zero CAM bilinear filter.

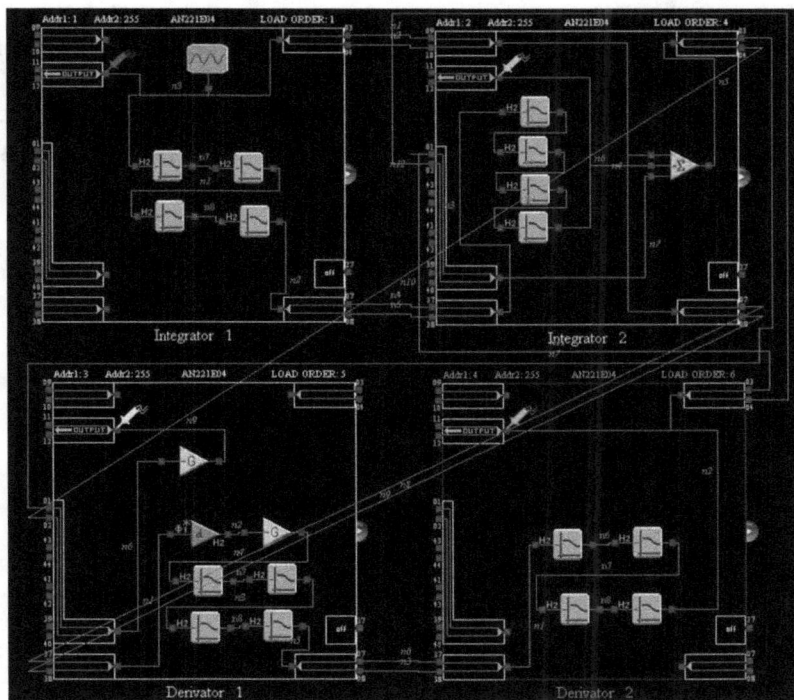

Fig. 6.3   FPAA schema of $PI^\lambda D^\mu$ controller.

The proportional coefficient has been obtained using an operational amplifier, the integral part has been implemented as previously described and the derivative part is obtained realizing the complement at one of the associated integral.

The number N of Bilinear Filter CAM used to realize the fractional integrator according the Oustaloup interpolation is 8 [Caponetto (2008a)].

The input signal, its integration, its derivative and output by $PI^\lambda D^\mu$ control are measured on-line real-time.

In Fig. 6.4, the bode diagram of an integrator of 0.7 order realized via Oustaloup interpolation is shown. This approximation is effective in a limited range of frequencies: $1 - 100\,rad/s$.

It must be noted that, the slope of the magnitude bode diagram is $-20mdb/dec$ (instead of $-20db/dec$ for first order system) and the phase angle approaches $-m\frac{\pi}{2}$ (instead of $-\frac{\pi}{2}$ for first order system).

It is quite evident that the fractional order m modulates the slope and the phase of the bode diagrams, providing a parameter that is useful for the open loop synthesis of the controller.

One important characteristic of FPAA is the on-fly parameters reconfiguration. This feature allows us to change the parameters dynamically without stopping the processing.

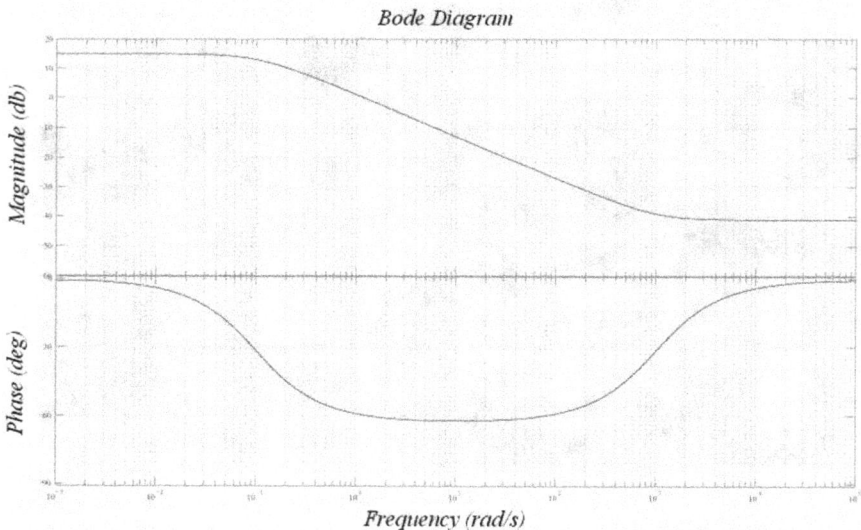

Fig. 6.4   Bode diagram of 0.7 integrator.

For each CAM a set of functions in C++ allows us to change the CAM parameters. In particular the parameters of the Bilinear Filter that are changed on-fly are the poles and zeros frequencies, and the static and high frequency gains.

6.2 Experimental Results

In Fig. 6.5 the software interface developed for on-fly parameters changing is reported. The interface provides four areas. The first is used to define the input parameters for simulation purposes. The remaining three areas are used to define the five parameters of the $PI^\lambda D^\mu$ controller.

In Figs. 6.6, 6.7, 6.8 and 6.9, a 0.5 gain effect, an integrative effect with $\lambda = 0.7$, a derivative action with $\mu = 0.2$, and the summed effects on the entire $PI^\lambda D^\mu$ controller have been reported.

In the first three examples the input is a sinusoidal signal with a frequency of $2Hz$ and a peak to peak amplitude $V_{pp} = 1.8V$, $V_{pp} = 1.4V$, $V_{pp} = 1.8V$, respectively.

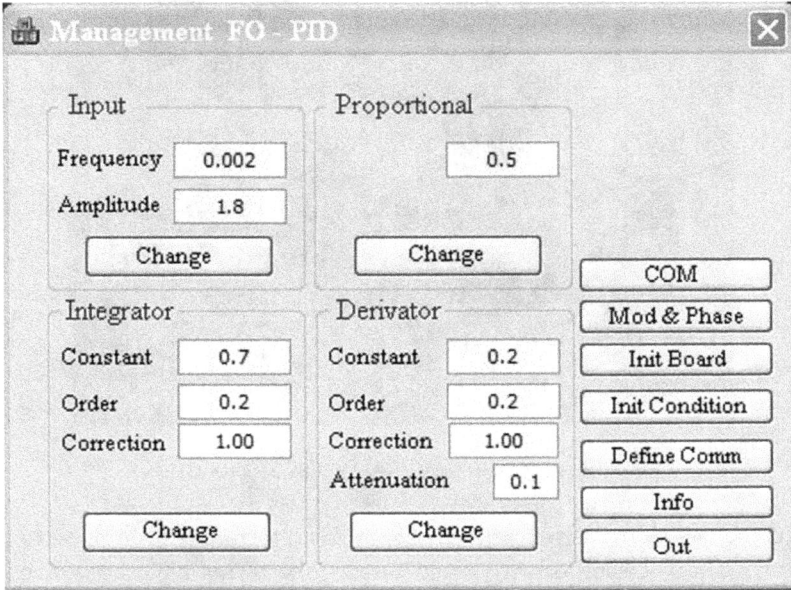

Fig. 6.5   Interface for changing on-line the parameters of $PI^\lambda D^\mu$.

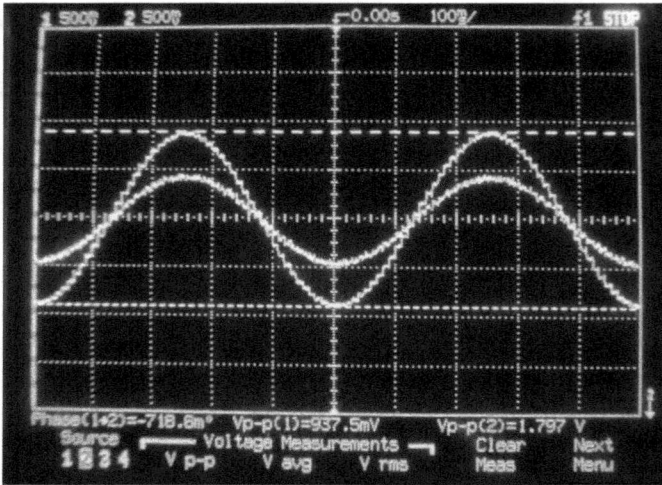

Fig. 6.6   Proportional effect of the $PI^\lambda D^\mu$ controller.

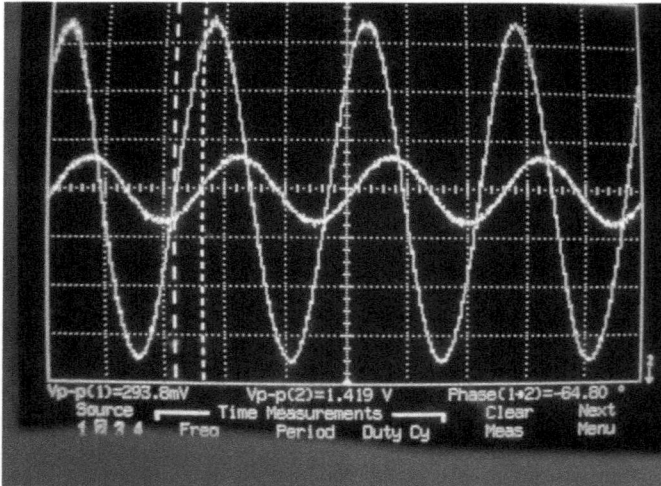

Fig. 6.7   Integrative effect of the $PI^\lambda D^\mu$ controller with $\lambda = 0.7$.

In the last example the gain of proportional term has been fixed to 0.5, the two gains of the derivative and integrative terms have been fixed to 0.2, while, the input is a sinusoidal signal with a peak to peak amplitude $V_{pp} = 1.8V$ and a frequency of $2Hz$.

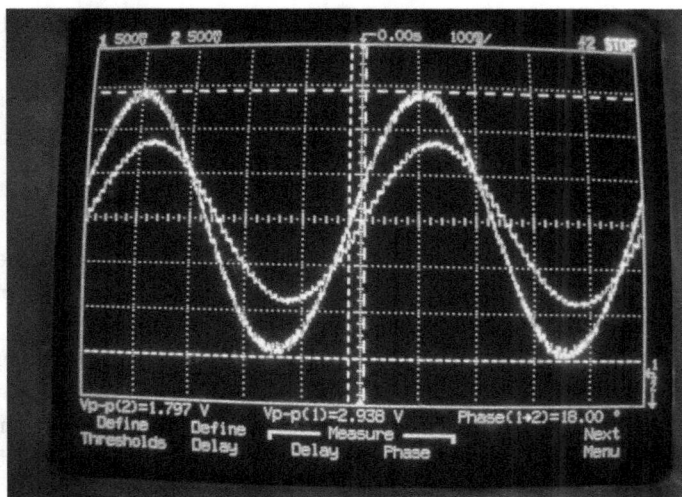

Fig. 6.8   Derivative effect of the $PI^\lambda D^\mu$ controller with $\mu = 0.2$.

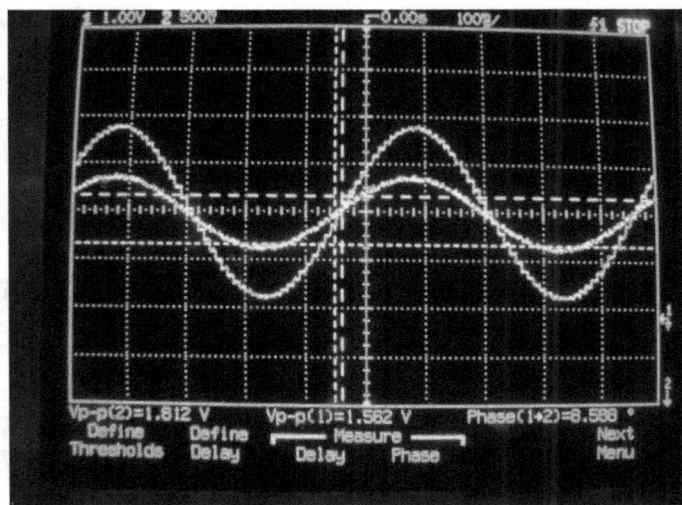

Fig. 6.9   Response of the $PI^\lambda D^\mu$, with $K_p = 0.5, K_i = K_d = 0.2, \lambda = 0.7, \mu = 0.2$ to a sinusoidal input with $V_{pp} = 1.8$ and a frequency of $2Hz$.

The signals reported in the Figs. 6.6 - 6.13 have been obtained using the "Hewlett Packard 54645D" oscilloscope. The sinusoidal input is connected on the channel 2 and the output on channel 1; the phase-difference between output and input is also shown.

In Fig. 6.6 the ideal values of the peak to peak output voltage and the phase-difference between input and output signals are respectively $900mV$ and $0°$.

The same signals in Figs. 6.7, 6.8 and 6.9 have the following values: $241mV$ and $-63°$, $2.98V$ and $18°$ and finally $1.51V$ and $5.8°$.

In all the figures the phase-difference is close to the ideal value, while for the output (channel 2) there is $50mV$ of overlapped noise. However this noise does not compromise the reliability of the developed system.

In Figs. 6.10 and 6.11, an integrative effect with $\lambda = 0.3$ and a derivative action with $\mu = 0.4$ have been reported. The two gains of the integrative and derivative terms have been fixed to 1, while, in this example, the input is a sinusoidal signal with a peak to peak amplitude $V_{pp} = 3.125V$ and $V_{pp} = 30.94mV$, respectively, and a frequency of $20Hz$.

In Fig. 6.10 the ideal values of the peak output voltage and the phase-difference are $733mV$ and $-27°$. In Fig. 6.11 they are $214mV$ and $36°$.

Also in these figures the phase-difference is close to the ideal value, while in output there is a small noise overlapped on the output voltage.

In Fig. 6.12, a derivative action with $\mu = 0.3$ is depicted. The gain of the derivative term has been fixed to 1, while, the input is a sinusoidal signal with a peak to peak amplitude $V_{pp} = 36.25mV$ and a frequency of $100Hz$.

In Fig. 6.12 the ideal values of the peak to peak output voltage and the phase-difference are $248mV$ and $27°$.

Modifying the poles and zeros frequencies of the Oustaloup interpolation it is possible to obtain the fractional action in another frequency range.

In Fig. 6.13, an integrative effect with $\lambda = 0.3$ is given. The gain of the integrative term has been fixed to 1, while, the input is a sinusoidal signal with a peak to peak amplitude $V_{pp} = 3.812V$ and a frequency of $5KHz$.

In Fig. 6.13 the ideal values of the peak to peak output voltage and phase-difference are $170mV$ and $-27°$.

Also in this case, at a higher frequency and using a different Oustaloup interpolation the proposed FPAA $PI^{\lambda}D^{\mu}$ controller continues to work correctly.

Finally the $PI^{\lambda}D^{\mu}$ FPAA implementation, which has been here discussed, can have relevant implications. Developments include the possibility of designing a dedicated IC, based on switch capacitors, as will be demonstrated in the next chapter, making it possible to produce a fine tuning adaptive non integer order controller.

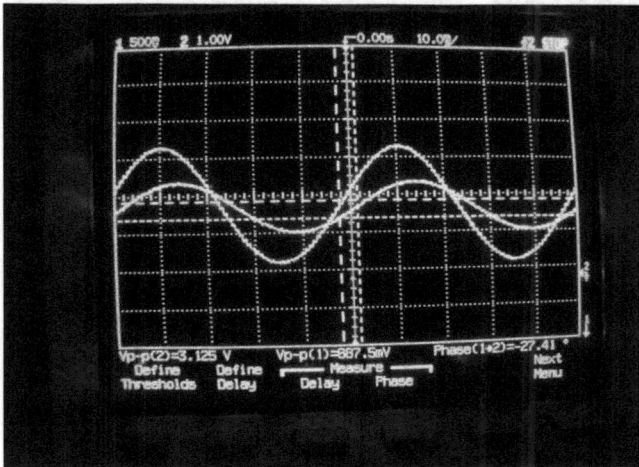

Fig. 6.10 Integrative effect of the $PI^\lambda D^\mu$ controller with $\lambda = 0.3$ at 20Hz.

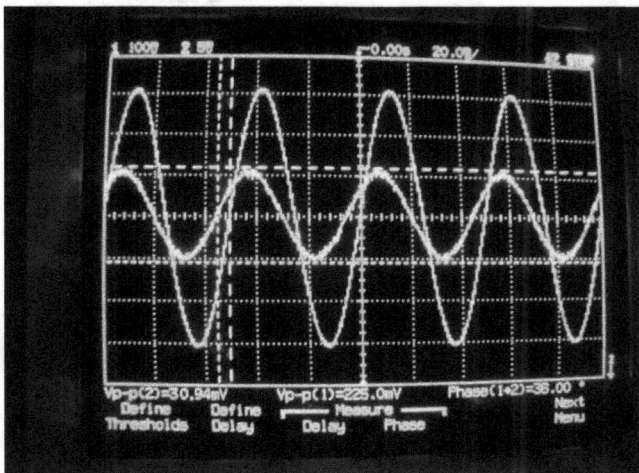

Fig. 6.11 Derivative effect of the $PI^\lambda D^\mu$ controller with $\mu = 0.4$ at 20Hz.

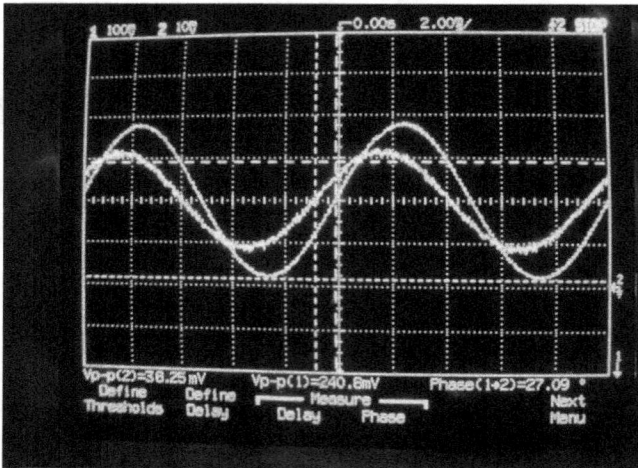

Fig. 6.12   Derivative effect of the $PI^\lambda D^\mu$ controller with $\mu = 0.3$ at 100Hz.

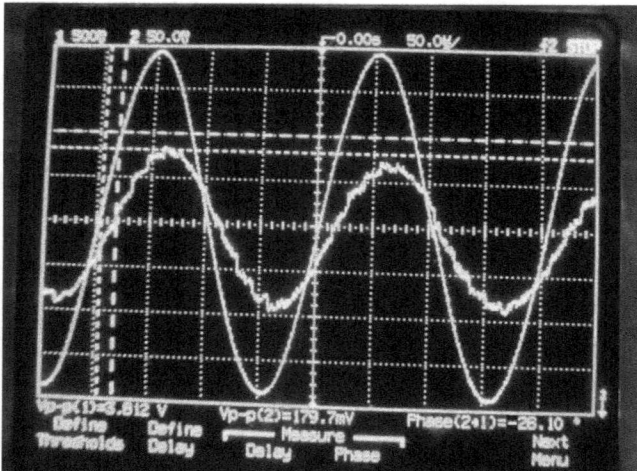

Fig. 6.13   Integrative effect of the $PI^\lambda D^\mu$ controller with $\lambda = 0.3$ at 5KHz.

Chapter 7

Switched Capacitor Integrated Circuit Design

A Switched Capacitors (SC) implementation of fractional differintegral operator is proposed in this chapter. Time and frequency domain result tests validate the feasibility and reliability of the SC circuit implementation. A detailed analysis of the influence of the non-idealities is proposed in order to obtain a design ready for Integrated Circuit (IC) implementation. The proposed approach may guarantee a good degree of approximation of the fractional differintegral operator and therefore may allow the possibility to realize a $PI^{\lambda}D^{\mu}$ controller IC based on switched capacitors technology.

7.1 Introduction

Ever since electric wave filters were introduced more than seventy years ago, filters have played a vital role in communication systems. With the advent of large scale integration (LSI) and very large scale integration (VLSI) technologies, greater emphasis has been placed on the miniaturization of the major components of communication systems, including filters. Consequently, one of the main directions of recent investigations into filters is the exploration of new technologies to achieve miniaturization.

The switched capacitor (SC) technique to design filters is a recent development and this is highly suitable for the creation of complete filters on a silicon chip. Switched capacitor filters realized using metal oxide semiconductor (MOS) technology use periodically operated switches with capacitors and operational amplifiers (OA). They are essentially based on active RC filter configurations, but they eliminate resistors through the use of switched capacitors. Furthermore, these are sampled data in nature and, therefore, require sampled data system theory for their analysis and design. Thus, SC filters are an outgrowth of active RC filter and digital filter theories.

Filters may also be classified according to the technology used to create them, or the application for which they are intended. We have thus a variety of filters, such as passive filters, active RC filters (which includes active R and active C filters), switched capacitor (SC) filters, and digital filters.

7.2   Passive and Active Filters

Passive filters use resistors, capacitors and inductors. Filters can be built using resistors and capacitors only. However, the resulting RC networks can realize only simple negative real-axis poles.

The use of inductors together with resistors and capacitors can result in network functions with complex-coniugate poles. Such RLC networks can create filters with a rapid variation of amplitude or phase response using a smaller number of elements (i.e., using a low-order filter) rather than an RC filter.

However, in practice it is preferable to avoid inductors , since they are bulky (especially if the inductance values are large) and non-ideal, and the realization of high-quality miniaturized inductors is not found to be practical.

It is possible to create resistances and capacitances in hybrid integrated circuits. Since RC networks by themselves cannot realize sharp-cutoff filters, highly selective amplitude or phase responses can be realized using passive RC networks along with active elements like bipolar transistors, operational amplifiers (OA), negative-impedance converters (NIC), gyrators, differential voltage-controlled current/voltage sources, or current conveyors.

However, the most popular among these is the OA. Active RC filters may use one or more OAs, together with passive RC networks, to realize the desired filtering functions. The OAs used in these active RC filters are assumed to have infinite input impedance, low output impedance and infinite d.c. gain.

Active RC filters, however, have some limitations, because of the non-ideal nature of the OAs used. Their performance is dependent on the finite bandwidth of OAs. ln addition, there will be variations due to temperature or power supply changes. The filter parameters are sensitive to resistor and capacitor values. These have to be considered in the design and implementation of active filters. In addition, designers are interested in considering the miniaturization of filters to an extent at which they are technologically compatible with other subsystems which use digital techniques.

These active RC filters can be realized using thin-film or thick-film or silicon bipolar integrated circuit (IC) technology. Each realization has its own advantages and disadvantages. It is not the intention of this study to discuss the various processes in detail (these are dealt with in the literature on active filters). It is sufficient to state that metal oxide semiconductor (MOS) technology has certain attractive features which have made it highly preferable to large scale and very large scale integration (LSI/VLSI) of electronic circuits.

7.3   Switched Capacitors Filters

These utilize a capacitor and two switches to simulate the circuit behaviour of a resistor, as shown in Fig. 7.1. The operation of the switched capacitor resistor is as follows. When the switch is in the left-hand position, C_1 is charged to voltage v_l and when the switch is thrown to the right, C_1 is discharged to voltage v_2. Thus, the amount of charge flowing into (or away from) v_2 is $Q_c = C_1(v_2 - v_1)$. On throwing the switch back and forth every T seconds, the current flow, i, into v_2 will be

$$i = \frac{C_1(v_2 - v_1)}{T} = \frac{(v_2 - v_1)}{T/C_1} \tag{7.1}$$

Thus, the switched capacitor simulates the behaviour of a resistor of value $(T/C_1) = R_1$, connected between voltage sources v_1 and v_2.

The circuit of Fig. 7.1 can be created in MOS technology using two MOS switches and a capacitor. If a capacitor C_2 is associated with the above SC resistor R_1, the resulting time constant $C_2 R_1$ is

$$\tau = C_2 R_1 = \left(\frac{C_2}{C_1}\right) T \tag{7.2}$$

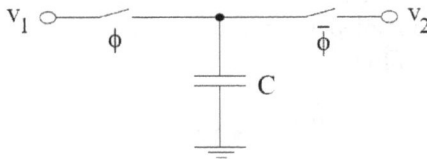

Fig. 7.1   A switched capacitor configuration simulates the function of a resistor.

Thus the time constant τ is determined by the ratio of the capacitors and not their absolute values, which makes it insensitive to process variations. This desirable property for integrated filter realizations, which avoids trimming, is achieved in the SC technique. It is thus seen that resistors in active RC filters can be realized by MOS switched capacitors. The availability of OAs in MOS technology can provide a solution to the realization of integrated filters [Hodges (1978)].

Commercially available OAs use largely bipolar technology and, in some cases, a combination of MOS and bipolar technologies. However,fully integrated MOS OAs have only become available recently, and these have boosted the possibility of the realisation of an SC filter in monolithic form.

7.4   Design of Sampled Data Filters

The design of sampled data filters starts with the specification of the desired frequency characteristics in the analog domain. The reason for such a choice is simple: there are considerable sources from which the design information in the analog domain can be obtained. Once the analog domain specifications and the transfer function specifying poles and zeros are obtained, the next step is to map these poles and zeros into the z-domain in such a way that the resulting sampled data filter satisfies the specifications with which the design procedure has been started.

The design of sampled data filters for given specifications in the analog domain will be illustrated in this section using different approaches: the impulse invariance method; the matched-z transformation method; the backward Euler integration (BEI) or p-transformation method; the forward Euler integration (FEI) method; the bilinear transformation (BT) method; and finally, the lossless discrete integrator (LDI) transformation technique.

7.4.1   *The Impulse Invariance Method*

This method ensures that the impulse response, $\hat{h}(n)$, of the sampled data filter, $H_d(s)$, is the sampled version of the impulse response, $h(t)$, of the corresponding analog filter, $H_a(s)$, by defining

$$\hat{h}(n) = h(t)|_{t=nT} \tag{7.3}$$

where T is the sample period.

Thus, it is defined as the *z-transformation* according to the impulse invariance method:

$$H_a(s) = \sum_{i=1}^{m} \frac{A_i}{s + s_i} \longrightarrow \sum_{i=1}^{m} \frac{A_i}{1 - e^{-s_i T} z^{-1}} \tag{7.4}$$

Although the time responses are essentially the same for analog as well as sampled data filters (due to the impulse invariance of the transform), the frequency responses will be different. This is due to the complex-plane mapping produced by the *z-transform*. The frequency response of the sampled data filter function is equal to the frequency response of the continuous function plus the contributions of the response displaced by multiples of $2\pi/T$. This addition or folding-in of these terms is called *aliasing*. The gain of an impulse invariant sampled data filter is proportional to the sampling frequency and may be considerably large.

It may be noted that the above method has been also designated by the standard z-transform method. It is applicable to the design of BP and all-pole LP filters. For other filter types, e.g., HP and BS, since a significant portion of the frequency response extends at high frequencies, the impulse invariance method maps the frequency response of the analog filter into the z-domain only up to $f_s/2$ and also the effect of aliasing is present. Hence, a useful alternative is the matched-z transformation method which will be considered next.

7.4.2   The Matched-z Transformation Method

In this method, the zeros and poles of the sampled data filter are matched to those of the continuous or analog filter through the relation:

$$z = e^{sT} \tag{7.5}$$

After transforming the poles and zeros, the gain, K, can be chosen to realize the desired gain for the sampled data transfer function:

$$H_d(z) = K \frac{\prod_{m=1}^{M}(1 - e^{-x_m T} z^{-1})}{\prod_{n=1}^{N}(1 - e^{-\alpha_n T} z^{-1})} \tag{7.6}$$

when the analog transfer function is given as

$$H_a(s) = \frac{\prod_{m=1}^{M}(s + x_m)}{\prod_{m=1}^{M}(s + \alpha_m)} \tag{7.7}$$

The poles of the resulting $H_d(z)$ are the same as those in the impulse invariance method, whereas the zeros are usually different.

However, the matched-z transformation, has certain limitations. When the zeros of an analog transfer function are at frequencies greater than half the sampling frequency, they will be aliased to a low frequency. A second case is the design of all-pole filters for which the sampled data versions obtained using the matched-z transformation technique do not adequately represent the desired continuous system.

The impulse invariance and matched-z transformations can be called "*exponential*" transformations since they are based on the exponential relationship given in eq. (7.5).

If these *z-transformations* are used the z-domain transcendent transfer function is obtained, i.e. it enables circuit implementations.

The algebraic substitution methods of sampled data filter design will be considered next. The first two methods to be discussed are based on the approximation of the derivatives by finite differences.

7.4.3   *Backward Euler Approximation of Derivatives*

In this method, the derivative is approximated by the first backward difference as follows:

$$\frac{dy}{dt}\bigg|_{t=nT} = \frac{y(nT) - y((n-1)T)}{T} \tag{7.8}$$

Similarly, for the second derivative d^2y/dt^2, using the same approximation, we have:

$$\frac{d^2y}{dt^2}\bigg|_{t=nT} = \frac{\left(\frac{dy}{dt}\right)\big|_{t=nT} - \left(\frac{dy}{dt}\right)\big|_{t=(n-1)T}}{T} \tag{7.9}$$

Substituting equation (7.8) in equation (7.9), and taking z-transforms, we obtain:

$$Z\left(\frac{dy}{dt}\bigg|_{t=nT}\right) = \left[\frac{1 - z^{-1}}{T}\right] Y(z) \tag{7.10}$$

and

$$Z\left(\frac{d^2y}{dt^2}\bigg|_{t=nT}\right) = \left[\frac{1 - z^{-1}}{T}\right]^2 Y(z) \tag{7.11}$$

In general, it follows that all the derivatives in the continuous-time transfer function can be approximated in a simple manner to yield the corresponding z-domain transfer function.

Thus the *Backward Euler z-transformation* is defined as:

$$s = \frac{1 - z^{-1}}{T} \qquad (7.12)$$

This transformation has also been named the *p-transformation* in literature.

Since the mapping provided by p-transformation yields stable sampled data filters corresponding to stable analog filters, it can be used to design such sampled data filters (see Fig. 7.2). However, it should be ensured that the obtained z-domain transfer function realizes the same poles and zeros as the s-domain filter. This is achieved by what is known as *pre-warping*, meaning that in order to obtain the s-domain poles (or zeros) at desired locations, we should choose the z-domain poles or zeros through the relationship $z = e^{sT}$.

Another transformation using the "forward difference" approximation for a derivative will be considered next.

7.4.4   *Forward Euler Approximation of Derivatives*

In this case, the derivative is approximated by the forward difference, i.e.,

$$\left. \frac{dy}{dt} \right|_{t=nT} = \frac{y((n+1)T) - y(nT)}{T} \qquad (7.13)$$

Following the same method as in the previous case, this corresponds to the transformation:

$$s \longrightarrow \frac{z - 1}{T} \qquad (7.14)$$

The resulting mapping is shown in Fig. 7.3. It is evident that stable analog transfer functions do not result in stable sampled data transfer functions when the forward Euler approximation is used.

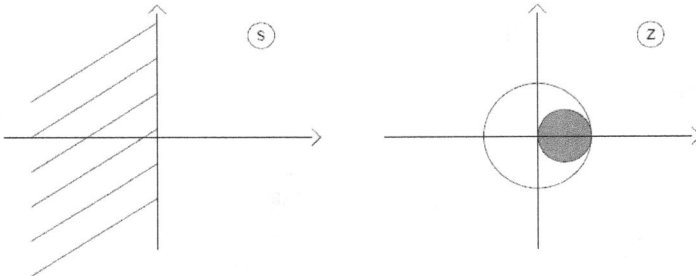

Fig. 7.2   Mapping resulting from backward Euler transformation of the derivative.

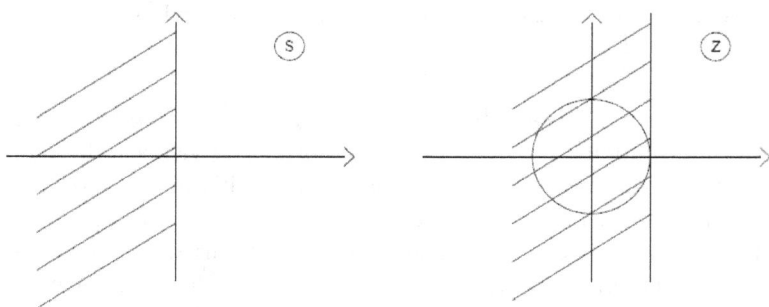

Fig. 7.3   Mapping resulting from forward Euler transformation of derivatives.

It is thus seen that the backward Euler approximation is preferable to the forward Euler approximation. Alternative procedures for obtaining sampled data transfer functions will be investigated next.

7.4.5   *The Bilinear Transformation Method*

The *bilinear z-transformation* is defined as:

$$s \longrightarrow \frac{2}{T}\left(\frac{1 - z^{-1}}{1 + z^{-1}}\right) \tag{7.15}$$

The *bilinear z-transformation* avoids the problem of aliasing encountered in the use of *impulse invariance* since it maps the entire imaginary axis in the s-plane onto the unit circle in the z-domain. However, the price paid for this is the introduction of the frequency warping effect or a distortion in the frequency axis. This can be corrected by "prewarping" the analog filter specifications.

It may be noted that while the bilinear transformation reproduces the amplitude response of the prototype analog filter in the amplitude response of the sampled data filter, it does not preserve the phase response. This is because of the distortion of the frequency axis. In the next subsection, another useful mapping transformation, the lossless discrete integrator (LDI) transformation, will be discussed.

7.4.6 *The Lossless Discrete Integrator Transformation*

For the LDI transformation, the derivative is approximated differently:

$$\frac{dy}{dt}\bigg|_{t=(n+1/2)T} = \frac{y(nT+T) - y(nT)}{T} \tag{7.16}$$

The first observation that can be made from this approximation is that the sampling points of the derivative and the original time function are not the same (when compared with the forward Euler or backward Euler approximations). The area under the curve dy_a/dt between sampling instants nT and $(n+l)T$ is approximated by midpoint integration.

From this definition the *LDI z-transformation* is defined as:

$$s \longrightarrow \frac{z^{1/2} - z^{-1/2}}{T} \tag{7.17}$$

The effect of transformation (7.17) is "warping" as discussed above for the bilinear transformation, but the "warping" for LDI and bilinear transformations is different in that bilinear transformation compresses the frequency scale whereas the LDI transformation expands it. Consequently, the frequency warping introduced by the bilinear transformation is more than that introduced by the LDI transformation.

7.5 Switched Capacitor Fundamental Circuits

The realization of Switched Capacitor circuits is not always possible using a single *z-rasformation*. There are parts of the circuit realized with one type of z-transformation, and other parts realized with other types of z-transformation. In particular a frequent situation is the presence of circuit blocks in a cascade realized by *Backward Euler transformation* and *Forward Euler transformation*. This configuration is equivalent to two blocks in a cascade realized by *LDI-transformation*.

It is now important to study a way how to transform a resistance in the continue time domain into switched capacitor configuration according to the z-transformations introduced in the previous chapter. Since there are more z-transformation definitions to move from the Laplace domain to the z domain it is natural that various ways exist how to implement a resistance with switched capacitor technology.

In order to realize the switched capacitors implementation of fractional differintegral operator, the Backward Euler transformation has been selected thanks to the characteristics descried previously.

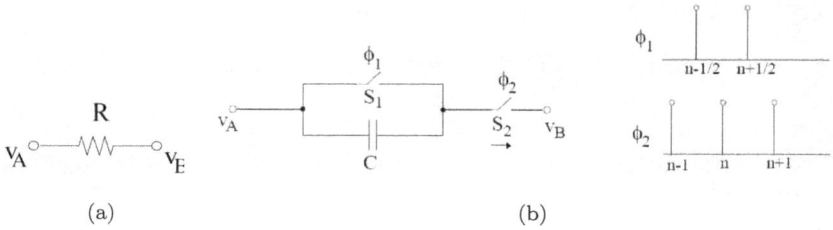

(a)　　　　　　　　　　　　　　(b)

Fig. 7.4　(a) Resistance, (b) Euler behind configuration.

7.5.1　*Resistor Realized by Backward Euler Transformation*

When using the resistance in Fig. 7.4(a), the current is $i = \frac{v_A - v_B}{R} = \frac{v_{AB}}{R}$.
But $i = \frac{dQ}{dt} \Rightarrow \frac{dQ}{dt} = \frac{v_{AB}}{R}$.

When applying the Laplace transformation:

$$i = \frac{Q(s)}{v_{AB}(s)} = \frac{1}{sR} \tag{7.18}$$

In order to obtain the transfer function of the switched capacitors configuration in Fig. 7.4(b) (Euler Behind Configuration) we must solve the equations of the circuit which are finite differential equations because the system is time-discrete.

Thus $Q(n) = Q(n-1) + C\big[v_A(n) - v_B(n)\big]$ because at the instant n, S_1 is open, S_2 is close and the capacitors discharge.

When applying the z-transformation $Q(z) = z^{-1}Q(z) + Cv_{AB}(z)$ or else:

$$H(z) = \frac{Q(z)}{v_{AB}(z)} = \frac{C}{1 - z^{-1}} \tag{7.19}$$

When using the Euler behind transformation for (7.18):

$$H(s)\Big|_{s=\frac{1}{T}\left(1-z^{-1}\right)} = \frac{T}{\left(1 - z^{-1}\right)} = \frac{\frac{T}{R}}{1 - z^{-1}} \tag{7.20}$$

When matching (7.19) and (7.20) the relation between the value of the switched capacitors and the value of the resistance is obtained:

$$C = \frac{T}{R} \tag{7.21}$$

7.6　Circuital Implementation of the Fractional Order Integrator

In this section a circuital implementation of Oustaloup interpolation of a fractional order integrator is proposed.

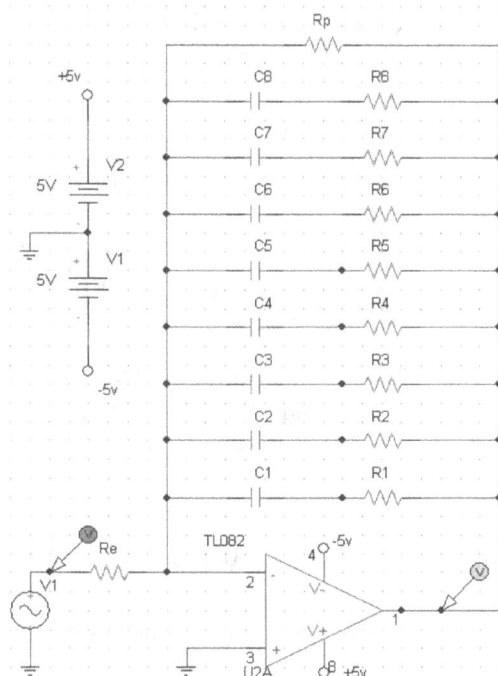

Fig. 7.5 Circuital implementation of Oustaloup interpolation of a fractional order integrator.

The transfer function in (1.44) can be realized, selecting appropriate values of the resistances and the capacities of the circuit shown in Fig. 7.5. In fact the transfer function of the circuit in Fig. 7.5 is:

$$H(jw) = \frac{R_p}{R_e} \prod_{i=1}^{N} (1 + jR_iC_iw)$$

$$* \frac{1}{\prod_{i=1}^{N}(1 + jR_iC_iw) + R_p \sum_{i=1}^{N} jC_iw \prod_{k=1, k \neq i}^{N}(1 + jR_kC_kw)} \quad (7.22)$$

Thus, matching (1.44) and (7.22) a set of equations is obtained, where in the first members are located the known terms joined to the fractional integration order and to the frequency range, according to the equations (1.45), while in the second members there are the circuit parameters need to be identified.

Thus, the following equations are obtained:

$$C_0 = \frac{R_p}{R_e} \tag{7.23}$$

$$w_i' = \frac{1}{R_i C_i} \qquad \forall i = 1 \ldots N \tag{7.24}$$

$$\prod_{i=1}^{N} 1 + j\left(\frac{w}{w_i}\right) = \prod_{i=1}^{N}(1 + jR_i C_i w) + R_p \sum_{i=1}^{N} jC_i w \prod_{k,k\neq i}^{N}(1 + jR_k C_k w) \tag{7.25}$$

An equations system of $2N + 1$ equations and $2N + 2$ parameters (R_p, R_e, R_i and $C_i \; \forall i = 1 \ldots N$)is obtained. It is therefore possible to fix one and calculate the others. A possible choice, suggested by empirical methods, is fix:

$$R_p = \frac{10^7}{W_u} \tag{7.26}$$

where $W_u = \sqrt{w_b w_h}$.

According to (7.24) we can rewrite (7.25) in the form:

$$\sum_{i=1}^{N} jC_i w \prod_{k=1,k\neq i}^{N} 1 + j\left(\frac{w}{w_k'}\right) = \frac{1}{R_p}\left[\prod_{i=1}^{N} 1 + j\left(\frac{w}{w_i}\right) - \prod_{i=1}^{N} 1 + j\left(\frac{w}{w_i'}\right)\right] \tag{7.27}$$

Equation (7.27) can be written in the following matrix form:

$$\left[jw\prod_{k=1,k\neq 1}^{N}(1+j\tfrac{w}{w_k'}) \quad \cdots \quad jw\prod_{k=1,k\neq N}^{N}(1+j\tfrac{w}{w_k'})\right] \begin{bmatrix} C_1 \\ \vdots \\ C_N \end{bmatrix}$$

$$= \frac{1}{R_p}\left[\prod_{i=1}^{N} 1 + j\left(\frac{w}{w_i}\right) - \prod_{i=1}^{N} 1 + j\left(\frac{w}{w_i'}\right)\right] \tag{7.28}$$

where the terms of the products are placed along the columns.

Thus there is a relation of the type $A * X = B \Leftrightarrow X = A^{-1} * B$ thanks to which it is possible to obtain the values of the capacities.

The values of the resistances are then obtained from the values of the capacities via the formula (7.24), in fact:

$$R_i = \frac{1}{w_i' C_i} \qquad \forall i = 1 \ldots N \tag{7.29}$$

7.7 Switched Capacitors Implementation of Fractional Order Integrator

In order to implement the fractional differintegral operator on chip, it is necessary that each resistance located in the circuit of Fig. 7.5 is replaced by a switched capacitors configuration [Caponetto (2008e)].

On replacing the resistances of the circuit in Fig. 7.5 with the switched capacitors calculated with (7.21) after selecting the switch time T of the switches, the circuit in Fig. 7.6 is obtained.

The capacitors used in the switched capacitors technology, in order to obtain a fine accuracy, are realized between two layers of poly-silicon. With these capacitors the minimal dimensions which can be obtained are $20 * 20 \mu m^2$ with values of capacities about $0.2 - 0.3 pF$. The capacities of greater value are realized connecting in parallel the capacities of the smaller value. This procedure allows to have greater tolerances to inaccuracies.

The switches are realized using a parallel configuration of n-MOS and p-MOS in a triode configuration in order to avoid the clock-feedthrough and increase the dynamics of the signals.

Finally, so as to define the values of the parameters of the circuit in Fig. 7.6, the value R_{on}, which characterizes the transistors in conduction, must be calculated. R_{on} is calculated according the time constant $\tau = R_{on}C$, where τ is the time of charge of a switch capacitor, as shown in Fig. 7.7. We must guarantee that the charge of the switched capacitor finishes before the end of the time slot, as shown in Fig. 7.7. Experimental results have revealed that the value:

$$R_{on}C \leq \frac{T}{14} \tag{7.30}$$

7.8 Results

The identification problem of the parameters of the circuit in Fig. 7.5, i.e the values of resistances and capacities calculated in (7.28) and (7.29), has been solved by developing a procedure in $Matlab^{®}$ Environment, which makes it possible to define the values of the resistances and capacities needed for the Oustaloup approximation of a fixed fractional order of integration on the condition that the range of frequency is fixed. The circuits are developed in $Orcad^{®}$ Environment.

In Table 7.1 the values of the resistances and capacities calculated by the previous procedure for a fractional integrator of order 0.5 are shown. In Fig. 7.8 the phase and module bode diagrams are plotted. In Figs. 7.9 and 7.10 the output waveform for a sinusoidal input of amplitude $1V$ and frequency $100Hz$ ($5Hz$ respectively) are shown. The amplitude of the

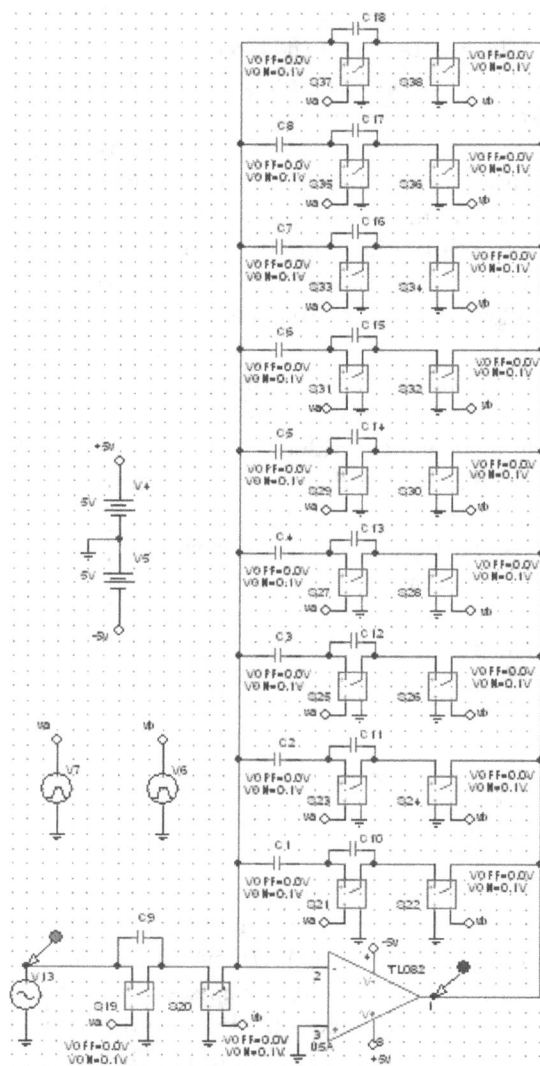

Fig. 7.6   Switched capacitors implementation of a fractional order integrator.

Table 7.1   Parameter values for a fractional integrator of order 0.5.

R_1	R_2	R_3	R_4
$2.825k\Omega$	$7.85k\Omega$	$15.29k\Omega$	$27.936k\Omega$

R_5	R_6	R_7	R_8
$50.294k\Omega$	$90.939k\Omega$	$169.71k\Omega$	$362.97k\Omega$

C_1	C_2	C_3	C_4
$75.12nF$	$85.49nF$	$138.8nF$	$240nF$

C_5	C_6	C_7	C_8
$422nF$	$738nF$	$1.25\mu F$	$1.84\mu F$

output signal, about $40mV$ ($178mV$ respectively), is correctly attenuated by a factor $\frac{1}{(2\pi f)^m}$, where f is the frequency of the input signal, and $m = 0.5$ is the order of integration. The phase of the output signal is delayed of $m * 90° = 45°$, i.e., the time delay between the signal output and signal input is $1.25ms$ ($25ms$ respectively), because the time-period of the input signal is $10ms$ ($200ms$ respectively).

In Table 7.2 the values of the resistances and capacities calculated by the previous procedure for a fractional integrator of order 0.2 are shown. In

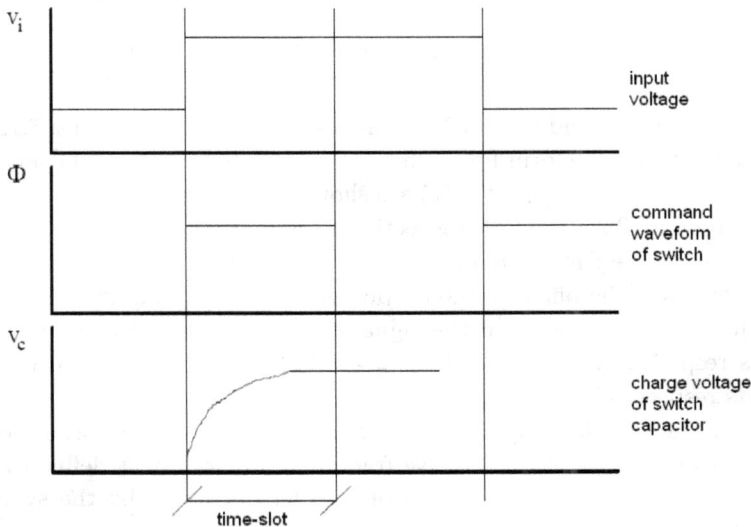

Fig. 7.7 R_{on} definition.

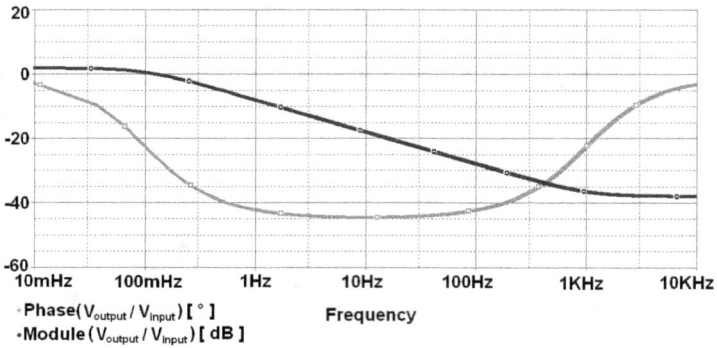

Fig. 7.8   Bode diagrams of fractional integrator of order 0.5.

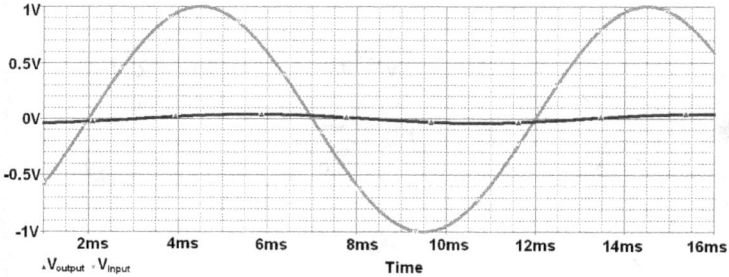

Fig. 7.9   Waveforms of an integrator of order 0.5 at $100Hz$.

Fig. 7.11 the phase and module bode diagrams are plotted. In Figs. 7.12 and 7.13 the output waveform for a sinusoidal input of amplitude $1V$ and frequency $100Hz$ ($5Hz$ respectively) are shown. The amplitude of the output signal, about $280mV$ ($500mV$ respectively), is correctly attenuated by a factor $\frac{1}{(2\pi f)^m}$, where f is the frequency of the input signal, and $m = 0.2$ is order of integration. The phase of the output signal is delayed of $m * 90° = 45°$, i.e., the time delay between the signal output and signal input is $0.5ms$ ($10ms$ respectively), because the time-period of the input signal is $10ms$ ($200ms$ respectively).

Using the switched capacitors implementation of the first order derivative operator and of the respective fractional integrator, as defined in eq. 1.6, it is possible to realize a fractional order derivative by the switched capacitors approach. The output waveform for a sinusoidal input of amplitude $100mV$ and frequency $5Hz$ is shown in Fig. 7.14.

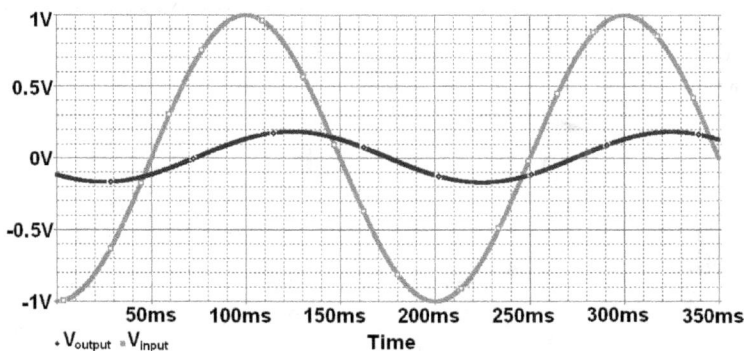

Fig. 7.10   Waveforms of an integrator of order 0.5 at $5Hz$.

Table 7.2   Parameter values for a fractional integrator of order 0.2.

R_1	R_2	R_3	R_4
$108.43k\Omega$	$155.09k\Omega$	$201.07k\Omega$	$255.46k\Omega$

R_5	R_6	R_7	R_8
$323.12k\Omega$	$409.85k\Omega$	$527.9k\Omega$	$727.86k\Omega$

C_1	C_2	C_3	C_4
$2.326nF$	$5.143nF$	$12.545nF$	$31.225nF$

C_5	C_6	C_7	C_8
$78.066nF$	$194.62nF$	$477.83nF$	$1.096\mu F$

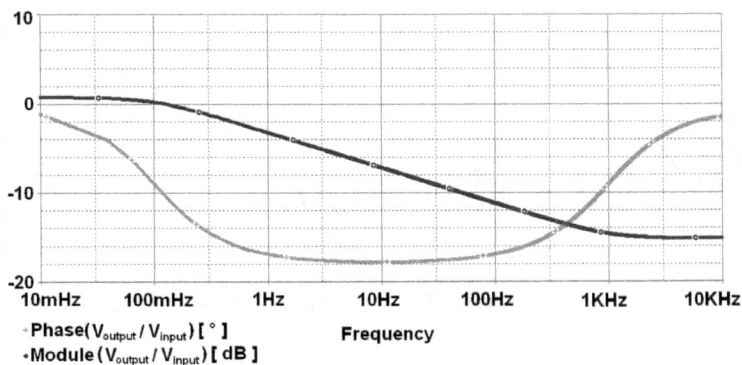

Fig. 7.11   Bode diagrams of fractional integrator of order 0.2.

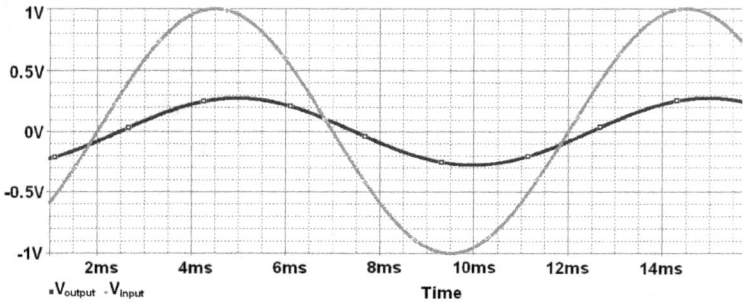

Fig. 7.12   Waveforms of an integrator of order 0.2 at $100Hz$.

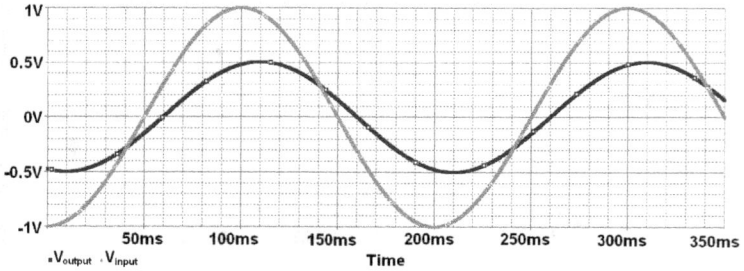

Fig. 7.13   Waveforms of an integrator of order 0.2 at $5Hz$.

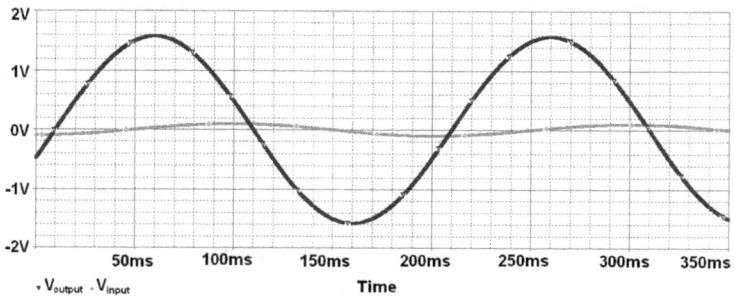

Fig. 7.14   Waveforms of a derivative operator of order 0.8 at $5Hz$.

Chapter 8

Fractional Order Model of IPMC

In this chapter a new model for Ionic Polymer Metal Composites (IPMC) actuators based on non-integer order models is proposed. IPMCs are very interesting polymers because of their capability to transform electrical energy into mechanical energy and vice-versa, making them particularly attractive for possible applications in different fields, such as robotics, aerospace, biomedicine, etc. An experimental setup has been realized to study the IPMCs behavior and an algorithm has been developed in Matlab environment in order to identify a fractional order model of IPMC actuators.

8.1 Fractional Model Identification Introduction

During the last century, the use of traditional differential calculus has become the fundamental tool to describe any physical phenomena that can be found in nature. A complex dynamic is usually modeled by a number N of first order differential equations. In systems theory, N is known as the degree of the system. Moreover, the theory of Laplace transformation in the case of linear systems has given the possibility of studying an input-output relation via the ratio of its Laplace transform, which is called the "Transfer Function" of the system and whose denominator is a polynomial of degree N. The duality of these approaches allows one to pass easily from the "Time Domain" of differential equations to the "Frequency Domain" of transfer functions. The reverse path is usually nontrivial and may cause some difficulties when dealing with classical mathematical tools. In fact there are physical phenomena whose study involves transfer functions of degree m, m being a non integer number. Transmission lines [Wang

(1987)], electrical noises [Mandelbrot (1967)], [Keshner (1982)], power-law [Korabel (2007)], control systems [Caponetto (2008a)], bioengineering [Margin (2006)], dielectric polarization [Onaral (1982)], ultracapacitor [Lorenzo (2008); Sabatier (2008)], heat transfer phenomena [LeMehaute (1991)], and systems with long-range interaction [Tarasov (2006, 2007)] are some of the fields having "Non Integer Order" physical laws.

For three centuries the theory of fractional derivatives developed mainly as a pure theoretical field of mathematics useful only for mathematicians. However, in the last few decades many authors pointed out that derivatives and integrals of non-integer order are very suitable for the description of properties of various real materials, e.g. polymers. It has been shown that new fractional-order models are more adequate than previously used integer-order models.

8.2　Ionic Polymer Metal Composites (IPMC)

IPMCs are innovative materials made of an ionic polymer membrane electroded on both sides with a noble metal. It is now well documented that IPMCs can exhibit large dynamic deformations if suitably electroded and forced by a time-varying voltage signal. Conversely, dynamic deformation of such ionic polymers produces dynamic electric fields across their electrodes. They can work either as low-voltage activated motion actuators or as motion sensors [Shahinpoor (2001)], [Shahinpoor (1998)]. These applications are shown in Fig. 8.1 and Fig. 8.2, respectively.

IPMCs show great potential as soft robotic actuators, artificial muscles and dynamic sensors in the micro-to-macro size range. They are ionic

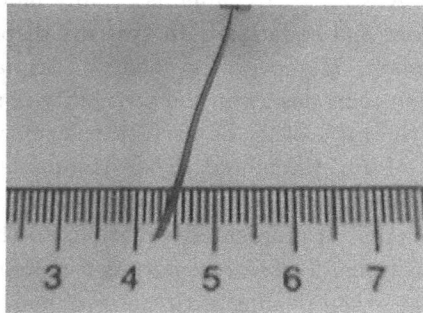

Fig. 8.1　Deformation of IPMC due to the applied voltage.

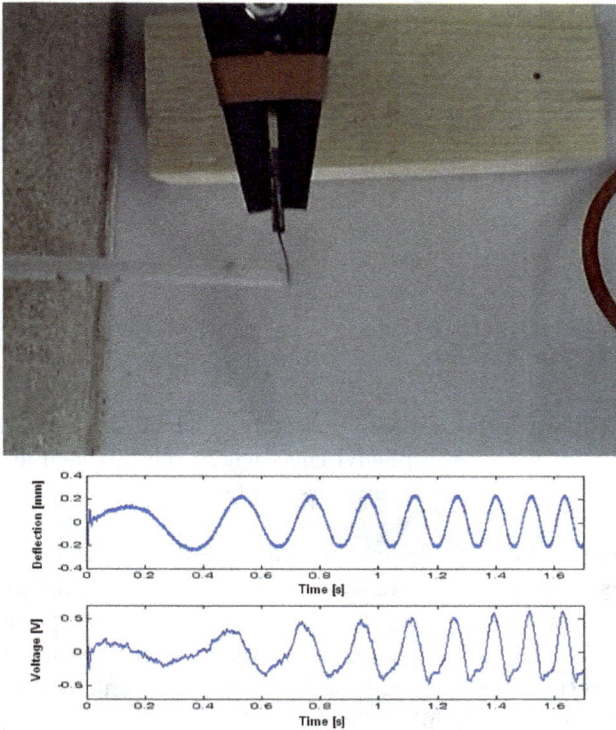

Fig. 8.2   IPMC used as a sensor in order to estimate the applied deformation.

polymers with ion exchanging capabilities which are then chemically treated
with an ionic salt solution of a metal and then chemically reduced to yield
ionic polymer metal composites. The term ion exchange polymers refers
to polymers designed to selectively exchange ions of a single charge (either
cations or anions) with their own incipient ions. They are often manufac-
tured from polymers that consist of fixed covalent ionic groups. Typical
ion exchange polymers are the following:

(1) Perfluorinated alkenes with short side-chains terminated by ionic
    groups (typically sulfonate or carboxylate (SO_3^- or COO^-) for cation
    exchange or ammonium cations for anion exchange (see Fig. 8.3)). The
    large polymer backbones determine their mechanical strength. Short
    side-chains provide ionic groups that interact with water and the pas-
    sage of appropriate ions.

$$-(CF_2CF_2)_n - \underset{\underset{CF_2}{|}}{CFO}(CF_2 - \underset{\underset{CF_3}{|}}{CFO})_m CF_2 CF_2 SO_3^- \cdots Na^+$$

or

$$-(CF_2CF_2)_x - \underset{\underset{CF_2}{|}}{CFO}(CF_2 - \underset{\underset{CF_3}{|}}{CFO})_m (CF_2)_n SO_3^- \cdots Na^+$$

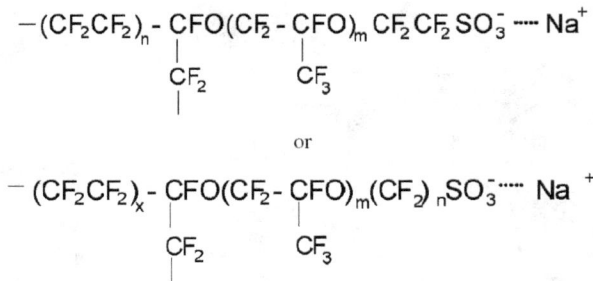

Fig. 8.3   Perfluorinated sulfonic acid polymers. The counter-ion, Na^+ in this case, can simply be replaced by other ions.

(2) Styrene/divinylbenzene-based polymers in which the ionic groups have been substituted by the phenyl rings where the nitrogen atom is fixed to an ionic group. These polymers are highly cross-linked and rigid.

In perfluorinated sulfonic acid polymers there are relatively few fixed ionic groups. They are located at the end of side-chains so as to position themselves in their preferred orientation to some extent. Therefore, they can create hydrophilic nano-channels, so-called cluster networks. Such configurations are completely different in other ionic polymers such as styrene/divinylbenzene families that are primarily limited by cross-linking, the ability of the ionic polymers to expand (due to their hydrophilic nature).

For the specific application the ionic polymer used is *Nafion* (produced by *Dupont* and distributed by *Sigma − Aldrich*), while Platinum has been chemically deposed to form the electrodes [Kim (2003)]. The chemical process is applied to large sheets of *Nafion*. Subsequently some sheets undergo an ionic exchange process where hydrogen ions are replaced by sodium or lithium ions (this step is necessary in order to improve mechanical transduction performance). Each sheet is then cut into strips, using a surgical blade, to obtain the Device Under Test (DUT). A photo of samples used as DUTs is shown in (Fig. 8.4).

The *Nafion* molecule has the structure given by the formula shown in Fig. 8.5.

Ionic polymers like *Nafion* have inner ionisable groups; a property of these groups is that they dissociate and move in the molecular net in a variety of solvent media. In the polymeric matrix of commercial Nafion-SO_3^- is the fixed group while the cation H^+ is free to move.

By applying a voltage to a typical electrolyte, cations and anions move

Fig. 8.4   A sample of IPMC.

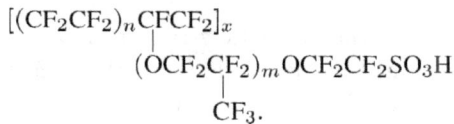

$$[(CF_2CF_2)_n \underset{|}{C}FCF_2]_x$$
$$(OCF_2\underset{|}{C}F_2)_m OCF_2CF_2SO_3H$$
$$CF_3.$$

Fig. 8.5   *Nafion* polymers.

in opposite directions: no energy is transferred from the molecular network to the solvent, and no solvent molecule is carried. Instead, in the above-mentioned polymeric membrane the solvent molecules can be carried parasitically by the mobile cations [Shahinpoor (1998)]. In the next section, a detailed description will be given of the IPMC material to which the innovative electric model proposed refers. Then it will be clear that the deformation of an IPMC actuator is strictly linked to the charge migration inside it. In order to predict IPMC deformation with a given input voltage, the first step is therefore to develop a model able to translate the input voltage into the relative current inside the membrane. This is the aim of this work.

8.3   Actuation Mechanism on IPMCs

When an external voltage is applied across the thickness of the IPMC, mobile cations will move toward the cathode. Moreover, if a hydrated sample is considered, the cations will carry water molecules with them. The cathode area will expand whilst the anode area will shrink; consequently the polymer will bend toward the anode. Cations with a high hydration number will produce a greater deformation than cations with a low hydration number [Shahinpoor (2001)]. For this reason, in motion-related applications, the hydrogen ion of the *Nafion* molecule in commercially available samples is purposefully substituted, via an ion exchange process, with Na^+, Li^+, etc.

A constant applied voltage causes the formation of a gradient of water concentration. Back diffusion of the water is responsible for the subsequent relaxation of the membrane [Nemat-Nasser (2002)].

Moreover, after deposition, metallic ions scatter throughout the superficial regions of the polymer, given its porous structure, where they are reduced to their metallic atoms. This results in the formation of dendritic electrodes, as shown in (Fig. 8.6). This phenomenon is widely reported in literature [Bar-Cohen (1999)], [Nemat-Nasser (2003)], [Bennett (2003)], where SEM images show the dendritic structures.

The two deposed metallic layers amplify the deformation: they add a deformation due to Coulombianforces to the effect of water transport. When subjected to an external voltage these electrodes will load one side with positive charges and the other with negative ones. These charges will interact with the fixed negative groups in the polymer chain. In the presence of an external electric field the positive ions move inside the polymer net, following the direction of the voltage gradient. Also, at the electrode-polymer interfaces the deformation phenomena are dominated by the interaction between the fixed negatively charged groups (e.g., SO_3^- in the case of *Nafion*) and the charged electrodes. The bending deformation induced by the metal is due to a differential contraction-expansion of the external fibers of the strips which are respectively subjected to attraction-repulsion forces between adjacent charges. The two actuation mechanisms described are represented in (Fig. 8.7).

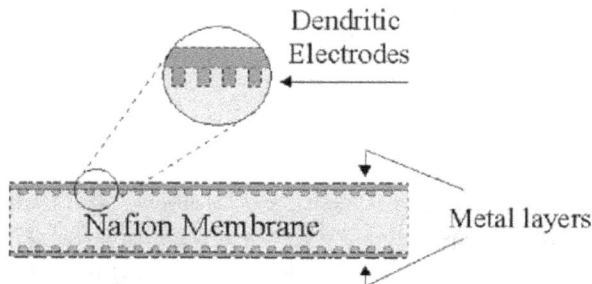

Fig. 8.6   Platinum distribution after deposition on the *Nafion* membrane surface.

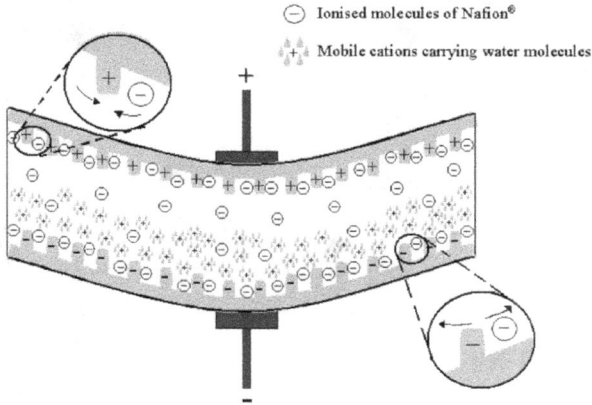

Fig. 8.7 Two actuation mechanisms inside an IPMC strip: water transportation and Coulombian forces due to the dendritic electrodes.

The bending of the IPMC membrane toward the anode is a function of the applied voltage up to a certain physical limit. Under ac voltage the membrane starts moving, following the applied electrical signal, while the nature of the deformation itself depends on both the amplitude and the frequency of the input wave [Shahinpoor (1999)]. More specifically, the displacement increases with amplitude while it decreases with frequency. The latter dependence can be accounted for by the inertia of the cations, which prevents them from following sudden variations in the applied voltage.

When a voltage signal is applied across the thickness of the IPMC, mobile cations will move toward the cathode. Moreover, if a hydrated sample is considered, the cations will carry water molecules with them. The cathode area will expand while the anode area will shrink; consequently the polymer will bend toward the anode (see Fig. 8.8(a)). Cations with a high hydration number will produce greater deformation than cations with a low hydration number [Nemat-Nasser (2002)]. For this reason, in motion-related applications, the hydrogen ion of the *Nafion* molecule is purposefully substituted, via an ion exchange process, by Na^+, Li^+, etc.

8.4 State-of-the-Art for IPMC Models

Several electrical models have been proposed in the literature in the last decade, but they only take the capacitive behavior of IPMCs into account,

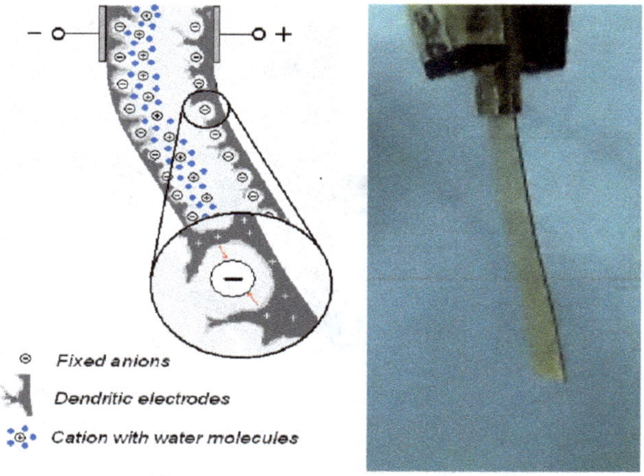

Fig. 8.8   (a) Chemical process and (b) beam of IPMC.

due to the presence of the polymeric membrane between two metal layers, and they do not give relevance to the cationic current term, which is the fundamental cause of the bending deformation. Also, all the models considered are linear, and therfore cannot describe the nonlinear phenomena of the material, which have been observed but have not been characterized yet. As a consequence, they are incomplete. Moreover, most of them use distributed circuits, which are not easy to analyze in order to find an analytical solution.

Considering the beam parameters, the length L_{free} and the cross-sectional dimensions (thickness t and width w), it will be assumed that the beam vibrates in the vertical plane (see Fig. 8.8(b)).

8.5   Experimental Setup

The experimental setup is composed of a circuit to impose the voltage input signal on the membrane and a distance laser sensor to measure the tip deflection. The scheme and the photo of the experimental setup are shown in Figs. 8.9 and 8.10, respectively.

When subjected to an external voltage, the IPMC dehydrates, and the resulting deformation decreases. Hence, the material must be re-hydrated to improve the repeatability of the observations. For this reason each

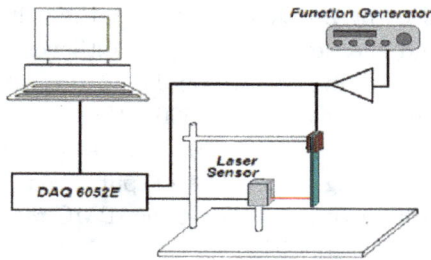

Fig. 8.9 The scheme of the measuring apparatus used to characterize the tip deflection of an IPMC strip.

Fig. 8.10 The photo of the measuring apparatus used to characterize the tip deflection of an IPMC strip.

acquisition survey was performed on a fully hydrated sample and, in order to minimize the hydration level dependence of the parameters, each acquisition survey lasted at most a few minutes. Moreover, after each acquisition, the IPMC strip was re-hydrated by immersion overnight in deionised water, until the next survey. Also, as long as the hydration level was restored, no significant degradation of the sample was observed by repeating several tens of surveys.

It can be noticed that the change in shape becomes significant from amplitudes higher than $1.5V$. For this reason a value of $1.5/2V$ was assumed as the threshold value.

The signals acquired by $DAQ6052E$, that is, the voltage input imposed on the membrane and the deflection of the cantilever tip measured with the laser sensor, are shown in Figs. 8.11 and 8.12, respectively. The voltage input signal is a linear chirp signal from 500 mHz to 50 Hz. Using a sample frequency equal to 1000 samples/s, 10000 samples are obtained for a data acquisition campaign lasting 10 s. The output signal acquired, i.e. the deflection of the cantilever tip, shows that the IPMC reaches the maximum deflection in the resonance condition

By processing these data in *Matlab* Environment, the transfer function of the system has been obtained, supposing that the system is linear, and using the *"tfestimate"* Matlab function. Thus, the Bode diagram of the system, shown in Fig. 8.13 has been estimated.

The frequency analysis of the IPMCs behaviour is limited to the range

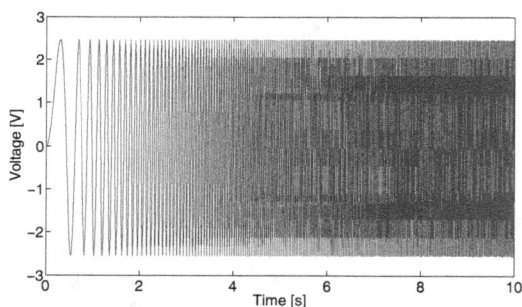

Fig. 8.11   Voltage input applied to the membrane.

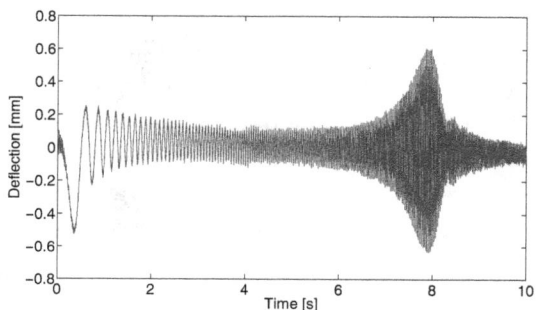

Fig. 8.12   Deflection of the cantilever tip measured with the laser sensor.

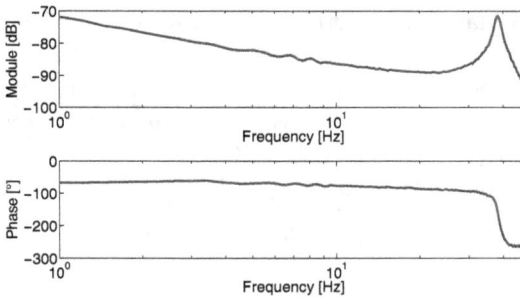

Fig. 8.13   Bode diagram of the voltage-deflection transfer function.

frequency of $0.5 - 50Hz$ because this is the range in which IPMCs work as actuators.

The inspection of the Bode diagrams makes it is clear that the system presents resonance at the frequency of about 40 Hz and, mostly, a non integer order behavior. In fact the initial slope in the module Bode diagram is about 15,5 db/dec, and the initial phase is about $-70°$ [Arena (2000)].

These Bode diagrams can also be identified by integer order models, but non integer order models make it possible to obtain a comparable modeling performance by using a smaller set of parameters [Caponetto (2008b)].

Therefore, it was decided that the model of IPMCs be identified with a non-integer order model. Since in this case the values of fractional exponents need to be estimated along with the corresponding transfer function zero and pole values, the identification problem is nonlinear and an adequate optimization procedure needs to be used.

8.6   Marquardt Algorithm for the Least Squares Estimation

Most algorithms for the least-squares estimation of nonlinear parameters have centered about either of two approaches. On one hand, the model may be expanded as a Taylor series with corrections applied to the several parameters calculated at each iteration on the assumption of local linearity. On the other hand, various modifications of the method of steepest-descent have been used. Both methods are not infrequently run aground: the Taylor series method because of the divergence of the successive iterations, the steepest-descent (or gradient) methods because of the agonizingly slow convergence after the first few iterations.

In this work a Marquardt method is used which, in effect, performs an optimum interpolation between the Taylor series method and the gradient method. This interpolation is based upon the maximum neighbourhood in which the truncated Taylor series gives an adequate representation of the nonlinear model [Marquardt (1963)].

Let the model to be fitted to the data be:

$$\hat{Y} = f(x_1, x_2, \ldots, x_m; \beta_1, \beta_2, \ldots, \beta_k) = f(\overrightarrow{x}, \overrightarrow{\beta}) \tag{8.1}$$

where $x_1, x_2, \ldots, x_m$ are independent variables, $\beta_1, \beta_2, \ldots, \beta_k$ are the population values of k parameters, and $\hat{Y}$ is the expected value of the function. Let the data points be denoted by $(Y_i, x_{1i}, x_{2i}, \ldots, x_{mi})$.

The problem is to compute those estimates of the parameters which will minimize

$$\Phi = \sum_{i=1}^{n} \left[Y_i - \hat{Y}_i \right]^2 = \left\| Y - \hat{Y} \right\|^2 \tag{8.2}$$

where $\hat{Y}_i$ is the value of Y predicted by (8.1) at the ith data point.

In the eq. (8.3) the Taylor series is written through the linear terms:

$$\left\langle Y\left(\overrightarrow{X_i}, \overrightarrow{b} + \overrightarrow{\delta_t}\right) \right\rangle = f\left(\overrightarrow{X_i}, \overrightarrow{b}\right) + \sum_{j=1}^{n} \left(\frac{\partial f_i}{\partial b_j} \right) (\delta_t)_j \tag{8.3}$$

or

$$\left\langle \overrightarrow{Y} \right\rangle = \overrightarrow{f_0} + P \overrightarrow{\delta_t} \tag{8.4}$$

In (8.3), $\overrightarrow{\beta}$ is replaced notationally by $\overrightarrow{b}$, the converged value of $\overrightarrow{b}$ being the least-squares estimate of $\overrightarrow{\beta}$. The vector $\overrightarrow{\delta_t}$ is a small correction to $\overrightarrow{b}$, with the subscript t used to designate $\overrightarrow{\delta}$ as calculated by this Taylor series method. The brackets $\langle \rangle$ are used to distinguish predictions based upon the linearized model from those based upon the actual nonlinear model.

Thus, the value of Φ predicted by (8.3) or (8.4) is

$$\left\langle \Phi \right\rangle = \sum_{i=1}^{n} \left[Y_i - \left\langle Y_i \right\rangle \right]^2 \tag{8.5}$$

Now, $\overrightarrow{\delta_t}$ appears linearly in (8.3) or (8.4), and can therefore be found by the standard least-squares method, by imposing the condition $\partial \langle \Phi \rangle / \partial \delta_j = 0$, for all j. Thus $\overrightarrow{\delta_t}$ is found by solving at r-th iteration the equation:

$$\left(A^{(r)} + \lambda^{(r)} I \right) \overrightarrow{\delta_t} = \overrightarrow{g}^{(r)} \tag{8.6}$$

where

$$A^{[k \times k]} = P^T P, \quad P^{[n \times k]} = \left(\frac{\partial f_i}{\partial b_j} \right),$$

$$g^{[k \times 1]} = \left(\sum_{i=1}^{n} (Y_i - f_i) \frac{\partial f_i}{\partial b_j} \right) = P^T (\overrightarrow{Y} - \overrightarrow{f_0}) \tag{8.7}$$

for $i = 1, 2, \ldots, n;\ j = 1, 2, \ldots, k$.

The new trial vector $b^{(r+1)} = b^{(r)} + \delta^{(r)}$ will lead to a new sum of squares $\Phi^{(r+1)}$. It is essential to $\lambda^{(r)}$ such that

$$\Phi^{(r+1)} < \Phi^{(r)} \tag{8.8}$$

It is clear from the foregoing theory that a sufficiently $\lambda^{(r)}$ always exists such that (8.8) will be satisfied, unless $\overrightarrow{b}^{(r)}$ is already at a minimum of Φ.

At each iteration it is desired to minimize Φ in the (approximately) maximum neighbourhood over which the linearized function will give adequate representation of the nonlinear function. Accordingly, the strategy to choose $\lambda^{(r)}$ must seek to use a small value of $\lambda^{(r)}$ when the guesses are in the immediate vicinity of the minimum. Large values of $\lambda^{(r)}$ should therefore be used only when necessary to satisfy (8.8). Such a strategy would inherit many of the properties of steepest-descent; e.g., rapid initial progress followed by progressively slower progress.

The strategy will therefore be defined as follows:

Let $\nu > 1$ be a real number.

Let $\lambda^{(r-1)}$ denote the value of λ from the previous iteration. Initially fixing $\lambda^{(0)} = 0.01$ $\Phi(\lambda^{(r-1)})$ is computed:

(1) If $\Phi(\lambda^{(r-1)}) > \Phi(\lambda^{(r)})$, let $\lambda^{(r)} = \lambda^{(r-1)}/\lambda$
(2) If $\Phi(\lambda^{(r-1)}) < \Phi(\lambda^{(r)})$, increase λ by successive multiplication by ν until for the some smallest w $\Phi(\lambda^{(r-1)}\nu^w) > \Phi(\lambda^{(r)})$. Let $\lambda^{(r)} = \lambda^{(r-1)}\nu^w$

By this algorithm a feasible neighbourhood is always obtained. Furthermore, the maximum neighborhood in which the Taylor series gives an adequate representation for our purposes is almost always obtained, within a factor determined by ν.

8.7   Fractional Models for IPMC Actuators

8.7.1   *Comparison Between an Integer Model and a Fractional Model of IPMC Actuators*

When applying the Marquardt algorithm to the available data, the model obtained is:

$$F(s) = \frac{340}{s^{0.756}(s^2 + 3.85s + 5880)^{1.15}} \qquad (8.9)$$

A comparison between the tip deflection of the cantilever IPMC as predicted by a model and corresponding acquired data is shown in Fig. 8.14 and Fig. 8.15. More specifically, the Module and Phase Bode diagrams are shown, respectively. The graph shown in the reported figures is referred to

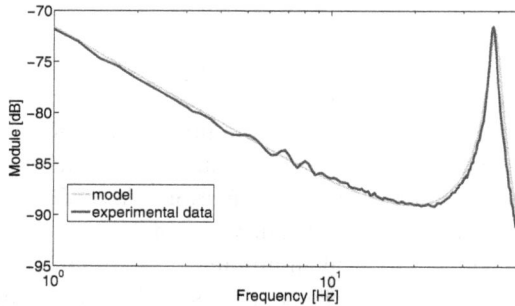

Fig. 8.14   Module comparison between predicted and measured tip deflection of a cantilever IPMC.

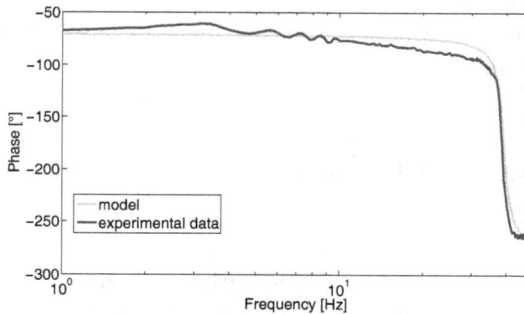

Fig. 8.15   Phase comparison between predicted and measured tip deflection of a cantilever IPMC.

as $117 Nafion$ IPMC with Sodium as a counter ion and $25mm$ long, $3mm$ wide, and $200\mu m$ thick. Results show a good prediction of the frequency response.

In Figs. 8.16 and 8.17 a comparison between the realized non-integer order model, the experimental acquired data and an integer order model, whose transfer function is explained in eq. (8.10), are shown.

$$F(s) = \frac{900s^2 + 2.57 * 10^6 s + 2.85 * 10^9}{0.011s^4 + 31.59s^3 + 4.97 * 10^2 s^2 + 2.57 * 10^6 s + 2.85 * 10^9} *$$

$$* \frac{3.47 * 10^{-3}s + 92.52 * 10^{-3}}{s^2 + 53.3s + 3.06 * 10^{-12}} \qquad (8.10)$$

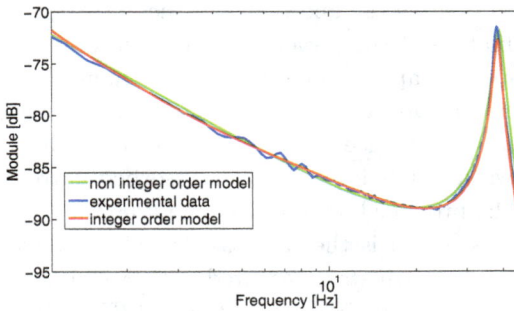

Fig. 8.16 Module comparison between the non-integer order model realized, the experimental acquired data and an integer order model.

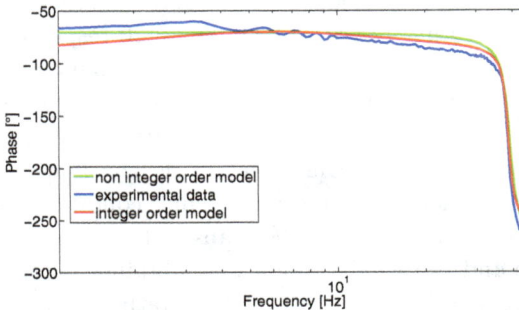

Fig. 8.17 Phase comparison between the non-integer order model realized, the experimental acquired data and an integer order model.

The same behavior of the non-integer order system identified can be approximated by a high integer order system. In particular the non-integer order model, proposed in this work, offers better modeling of the IPMC actuators at low frequencies.

8.7.2 *Fractional Models for the Electrical and Electromechanical Stages of IPMC Actuators*

In this section new models for the electrical and electromechanical stages of IPMC actuators, based on non-integer order models, are proposed [Caponetto (2008d)].

The new models of the electrical and electromechanical stages of IPMC actuators allow us to estimate the IPMC actuator absorbed current and the relevant mechanical quantities (free deflection and/or blocked force). A key factor in actuator design is the knowledge of the absorbed current; indeed, it allows the estimation of the power consumption for the device. This is a fundamental aspect in applications such as autonomous robotic systems, where the designer must address the problem of power source availability. It is therefore of great importance to have a single model able to describe the relationships between applied voltage and absorbed current and between this current and the produced action [Bonomo (2007)].

The experimental setup is the same as described previously. The deflection of the cantilever tip was measured with a commercially available distance laser sensor (*Baumer Electrics OADM12U6430*). Light from the laser diode was focused onto the end of the cantilever. The absorbed current was transduced by using a shunt resistor. The signals acquired by *DAQ6052E*, that is, the voltage input imposed on the membrane, the current absorbed, and the deflection of the cantilever tip, measured with the laser sensor, are shown in Figs. 8.18, 8.19, and 8.20 respectively.

The voltage input signal is a linear chirp signal from $500mHz$ to $50Hz$. Using a sample frequency equal to 1000 samples/s, 10000 samples are obtained for a data acquisition campaign during 10 s. The acquired output signal, i.e. the deflection of the cantilever tip, clearly shows that the IPMC reaches the maximum deflection in the resonance condition. On processing this data in *Matlab* Environment, the transfer functions voltage-current, current-deflection and voltage deflection were obtained, supposing that the system is linear, and using the "tfestimate" Matlab function. Thus, the Bode diagram of the three functions, shown in Figs. 8.21, 8.22, and 8.23, have been estimated.

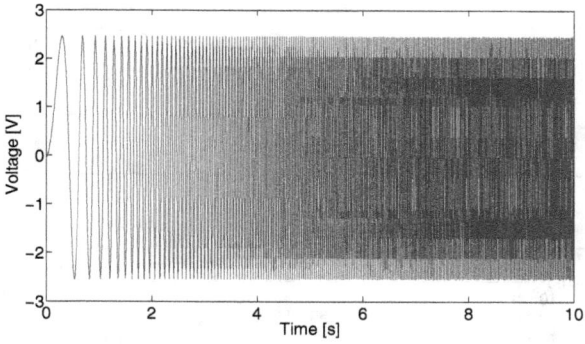

Fig. 8.18   Voltage input applied to the membrane.

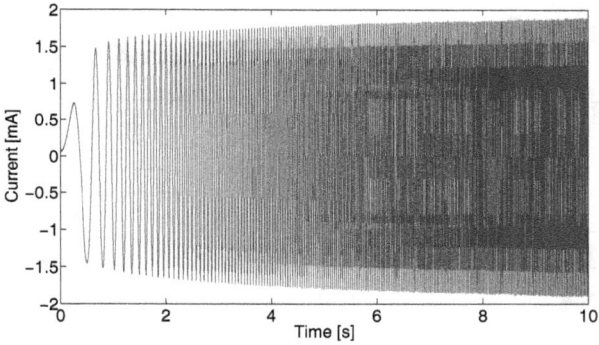

Fig. 8.19   Current absorbed by the membrane.

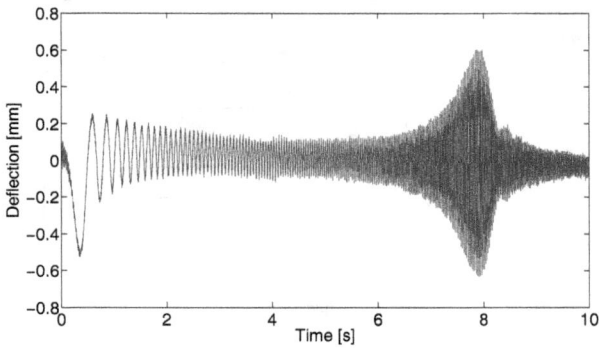

Fig. 8.20   Deflection of the cantilever tip measured with the laser sensor.

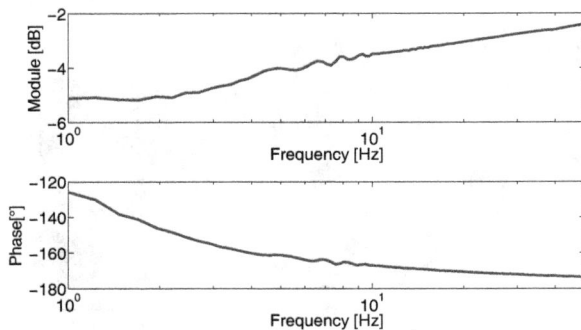

Fig. 8.21   Bode diagrams of the system voltage-current deduced from experimental data.

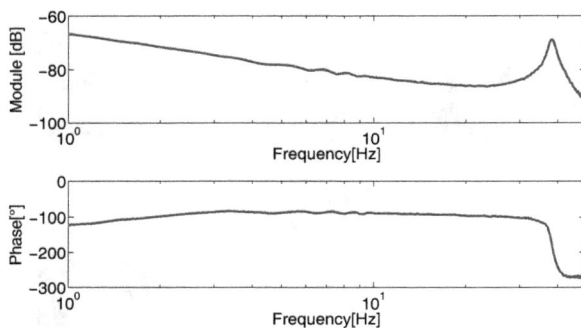

Fig. 8.22   Bode diagrams of the system current-deflection deduced from experimental data.

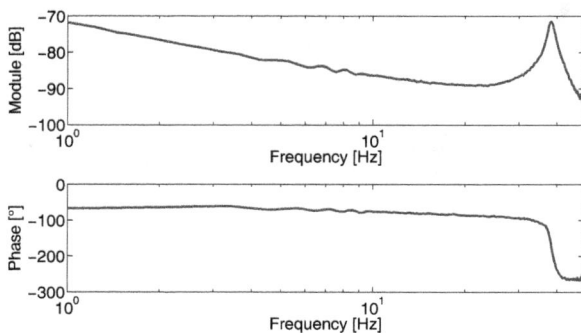

Fig. 8.23   Bode diagrams of the system voltage-deflection deduced from experimental data.

On inspecting the Bode diagrams it becomes clear that the three systems present a non integer order behaviour [Arena (2000)]. In fact the module Bode diagrams present a slope equal to $m * 20\ db/decade$, and the phase Bode diagrams present a phase lag equal to $n * 90°$, where m and n are real numbers.

As previously it was decided to identify the models of the three systems with non-integer order models. Since in this case the values of fractional exponents need to be estimated along with the corresponding transfer function zero and pole values, the identification problem is nonlinear and the *Marquardt Algorithm* is used.

Applying this algorithm to the available data, the models obtained for the three transfer functions, voltage-current, current-deflection and voltage-deflection are shown in eqs. (8.11), (8.12), and (8.13) respectively.

$$\frac{I(s)}{V(s)} = 0.5s^{0.09}\frac{\left(\frac{s}{0.01}+1\right)^{1.2}}{\left(\frac{s}{1.5}+1\right)^{1.2}} \tag{8.11}$$

$$\frac{D(s)}{I(s)} = \frac{680}{s^{0.876}(s^2+3.85s+5880)^{1.15}}\frac{\left(\frac{s}{1.5}+1\right)^{1.2}}{\left(\frac{s}{0.01}+1\right)^{1.2}} \tag{8.12}$$

$$\frac{D(s)}{V(s)} = \frac{340}{s^{0.756}(s^2+3.85s+5880)^{1.15}} \tag{8.13}$$

A comparison between the transfer functions as predicted by the model and corresponding acquired data is shown in Figs. 8.24–8.29. More specifically, the Module and Phase Bode diagrams are shown, respectively. The graph shown in the reported figures is referred to a $117Nafion$ IPMC with Sodium as counter ion and $25mm$ long, $3mm$ wide, and $200\mu m$ tick. Results show a good prediction of the frequency response.

Another important comment regarding the three transfer functions (8.11), (8.12), and (8.13) is that in the voltage-current and current-deflection transfer functions there are dynamics that mutually compensate them and do not appear in the total voltage-deflection transfer function.

In order to validate the obtained models also in the time domain, the estimated value of the deflection of the cantilever tip by the non integer order model is plotted in Fig. 8.30. This faithfully follows the values measured with the laser sensor.

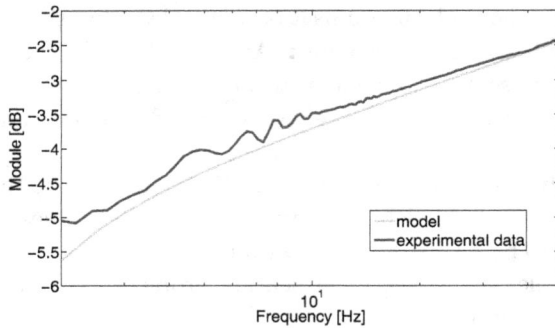

Fig. 8.24   Module comparison between predicted and measured voltage-current transfer function.

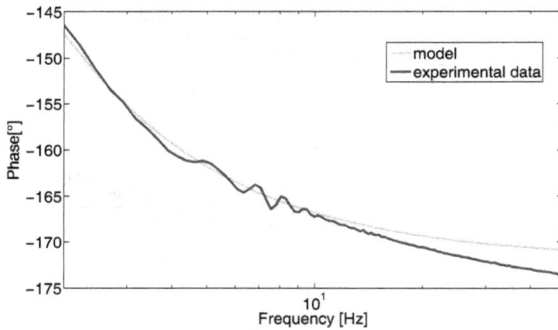

Fig. 8.25   Phase comparison between predicted and measured voltage-current transfer function.

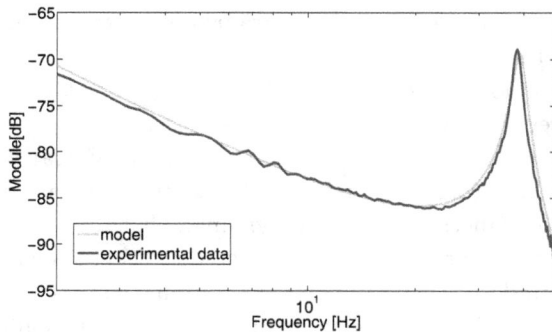

Fig. 8.26   Module comparison between predicted and measured current-deflection transfer function.

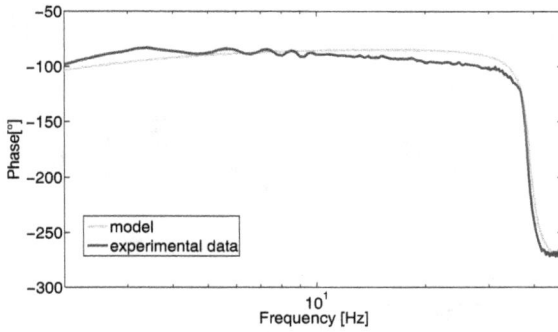

Fig. 8.27 Phase comparison between predicted and measured current-deflection transfer function.

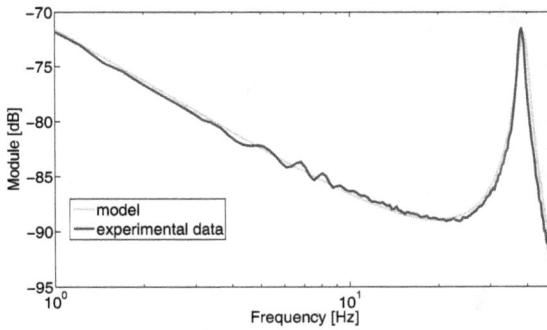

Fig. 8.28 Module comparison between predicted and measured voltage-deflection transfer function.

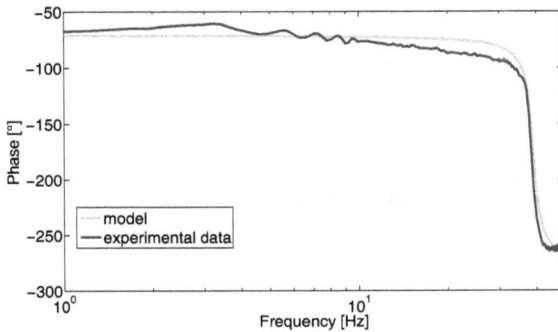

Fig. 8.29 Phase comparison between predicted and measured voltage-deflection transfer function.

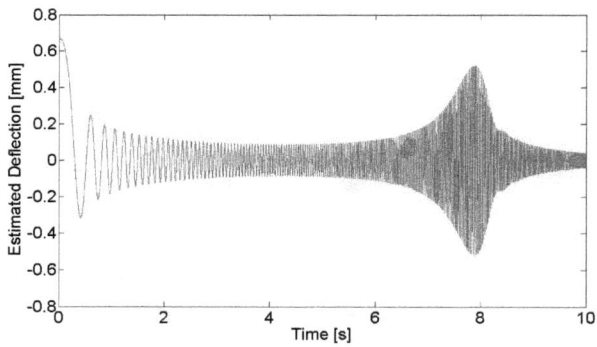

Fig. 8.30 Comparison in the time domain between the deflection measured with the laser sensor, and the value estimated by the non integer order model.

Bibliography

Ahmed, E., El-Sayed, A. M. A., and Hala El-Saka, A. A. (2007). *Equilibrium points, stability and numerical solutions of fractional-order predatorprey and rabies models, J. Math. Anal. Appl.*, vol. 325, pp. 542–553.

Akcay, H., and Malti, R. (2008). On the Completeness Problem for Fractional Rationals with Incommensurable Differentiation Orders, in: *Proc. of the 17th World Congress IFAC* (Soul, Korea), July 6-11, pp. 15367–15371.

Al-Alaoui, M. A. (1993). Novel digital integrator and differentiator, *Electron. Lett.*, vol. 29, no. 4, pp. 376-378.

Al-Alaoui, M. A. (1997). Filling the gap between the bilinear and the backward difference transforms: An interactive design approach, *Int. J. Elect. Eng. Edu.*, vol. 34, no. 4, pp. 331-337.

Anderson, B. D. O., Bose, N. I., and Jury, E. I. (1974). A Simple Test for Zeros of a Complex Polynomial in a Sector, *IEEE Transactions on Automatic Control*, Tech. Notes & Corresp., vol. AC-19, no. 4, pp. 437-438.

Aoun, M., Malti, R., Levron F., and Oustaloup A. (2007). Synthesis of fractional Laguerre basis for system approximation, *Automatica*, **43**, pp. 1640–1648.

Aoun, M., Malti, R., Levron F., and Oustaloup, A. (2004). Numerical Simulations of Fractional Systems: An Overview of Existing Methods and Improvements, *Nonlinear Dynamics*, **38**, pp. 117-131.

Arena, P., Bertucco, L., Fortuna, L., Nunnari, G., and Porto, D. (1998). CNN with Non-integer order cells, in *IEEE International Workshop on Cellular Neural Networks* (CNNA'98) (London, Great Britain), pp. 372–378.

Arena, P., Caponetto, R., Fortuna, L., and Porto, D. (2000). *Non Linear Non Integer Order Systems - An Introduction* (World Scientific).

Astrom, K. J (2000). *Model Uncertainty and Robust Control*, COSY project.

Bar-Cohen, Y., Leary, S., Yavrouian, A., Oguro, K., Tadokoro, S., Harrison, J., Smith, J., and Su, J. (1999). Challenges to the transition of IPMC artificial muscle actuators to practical applications, in *Proc. MRS Symposium*, Nov. 29 - Dec. 1, Boston, pp. 1-8.

Barbosa, R. S., Machado, T. J. A., Vinagre, B. M., and Calderón, A. J. (2007). *Analysis of the Van der Pol Oscillator Containing Derivatives of Fractional Order, Journal of Vibration and Control*, vol. 13, no. 9-10, pp. 1291–1301.

Bayat, F. M., and Afshar, M. (2008). Extending the Root-Locus Method to Fractional-Order Systems, *Journal of Applied Mathematics*, **528934**,13 DOI:10.1155/2008/528934.

Bayat, F. M., Afshar, M., and Ghartemani, M. K. (2009). Extension of the root-locus method to a certain class of fractional-order systems, *ISA Transactions*, vol. 48, no. 1, pp. 48–53.

Bellman, R., and Cooke, K. L. (1963). *Differential-Difference Equations* (Academic).

Bennett, M. D., and Leo, D. J. (2003). Manufacture and characterization of ionic polymer transducers employing, *Smart Materials and Structure*, **12**, pp. 424-436.

Bhattacharyya, S.P., Silva, G.J., and Datta, A. (2002). New Results on the Synthesis of PID Controllers, *IEEE Trans. Automatic Control*, **47**, pp. 241–252.

Bhattacharyya, S. P., Datta, A., and Ho, M. T. (2000). *Structure and Synthesis of PID Controller* (Springer-Verlag).

Bhattacharyya, S. P., Chapellat, H., and Keel L. H. (1995). *Robust Control: The Parametric Approach* (Prentice-Hall).

Bode, H. W. (1949). *Network Analysis and Feedback Amplifier Design* (Tung Hwa Book Company).

Bohannan, G. W. (2006). Analog Fractional Order Controller in a Temperature Control Application, in *Proc. IFAC Workshop on Fractional Differentiation and its Application* (FDA'06) (Porto, Portugal).

Bonomo, C., Fortuna, L., Giannone, P., Graziani, S., and Strazzeri, S. (2007). A nonlinear model for ionic polymer metal composites as actuators, *Smart Material and Structures*, **16**, pp. 1–12.

Caponetto, R., Dongola, G., Fortuna, L. (2007). A New Class of Fault-Tolerant Systems: FPGA Implementation of Bio-Inspired Self-Repairing System, in *Proc. 15th Mediterranean Conference on Control and Automation* (MED '07) (Athens, Greece).

Caponetto, R., and Dongola, G. (2007). Analog Implementation of Non Integer Order $PI^\lambda D^\mu$ controller via Field Programmable Analog Array, in Proc *ASME International Design Engineering Technical Conferences* (IDETC'07) (Las Vegas, Nevada).

Caponetto, R., and Dongola, G. (2008). Field Programmable Analog Array Implementation of Non Integer Order $PI^\lambda D^\mu$ Controller *Journal of Computational and Nonlinear Dynamics*, **3**, 021302.

Caponetto, R., Dongola, G., Fortuna, L., Graziani, S., and Strazzeri, S. (2008). A Fractional Model for IPMC Actuators, in *Proc. IEEE International Instrumentation and Measurement Technology Conference* (I^2MTC2008) (Vancouver Island, Canada).

Caponetto, R. and Dongola, G. (2008). New Algorithms for the Synthesis of Controller, in *Pr0c. IFAC Fractional Differentiation and its Applications* (FDA'08) (Ankara, Turkey).

Caponetto, R., Dongola, G. (2008). Fractional Models for the Electrical and Electromechanical Stages of IPMC Actuators, in *Proc. IFAC Fractional Differentiation and its Applications* (FDA'08) (Ankara, Turkey).

Caponetto, R. and Dongola, G. (2008). Switched Capacitors Implementation of Non Integer Order $PI^\lambda D^\mu$ Controller, in *Proc. IFAC Fractional Differentiation and its Applications* (FDA'08) (Ankara, Turkey).

Caponetto, R., and Dongola, G. (2006). Analog implementation of non integer order integrator via field Programmable Analogic array, in *Proc. IFAC Workshop on Fractional Differentiation and its Application* (FDA'06) (Porto, Portugal).

Caponetto, R., Fortuna, L., and Porto, D. (2004). A new tuning strategy for non integer order PID controller, in *Proc IFAC Workshop on Fractional Differentiation and its Application* (FDA'04) (Bordeaux, France).

Carlson, G. E., and Halijak, C. A. (1964). Approximation of fractional capacitors $(1/s)^{1/n}$ by a regular Newton process, *IEEE Transaction on Circuit Theory*, **11**, 2, pp. 210–213.

Charef, A. (2006). Modeling and Analog Realization of the Fundamental Linear Fractional Order Differential Equation, *Nonlinear Dynamics*, **46**, pp. 195-210.

Chen, Y. Q., Ahn, H.-S., and Xue D. (2006). Robust controllability of interval fractional order linear time invariant systems, *Signal Processing*, **86**, pp. 2794-2802.

Chen, Y. Q. (2003). Oustaloup Recursive Approximation for Fractional Order Differentiators, www.mathworks.com/matlabcentral/fileexchange/3802.

Chen, Y. Q., and Moore, K. L. (2002). Discretization schemes for fractional-order differentiators and integrators. *IEEE Trans. On Circuits and Systems - I: Fundamental Theory and Applications*, vol. 49, no. 3, pp. 363–367.

Das, S. (2007). Functional Fractional Calculus for System Identification and Controls (Springer).

Deng, W. (2007). Short memory principle and a predictor-corrector approach for fractional differentional equations, *Journal of Computational and Applied Mathematics*, **206**, pp. 174–188.

Deng, W. (2007). Numerical algorithm for the time fractional Fokker-Planck equation, *Journal of Computational Physics*, **227**, pp. 1510–1522.

Deng, W., Li Ch., and Lu, J. (2007). Stability analysis of linear fractional differential system with multiple time delays, *Nonlinear Dyn.*, f48, pp. 409–416.

Deng, W. H., and Li, C. P. (2005). *Chaos synchronization of the fractional Lu system, Physica A*, vol. 353, pp. 61–72.

Deregel, P. (1993). *Chua's oscillator: Z zoo of attractors, Journal of Circuits, Systems, and Computers* , vol. 3, no. 2, pp. 309–359.

Diethelm, K., Ford, N. J., Freed, A. D., and Luchko, Yu, (2005). Algorithms for the fractional calculus: A selection of numerical methods, *Comput. Methods Appl. Mech. Engrg.*, **194**, pp. 743–773.

Dorčák, Ľ. (1994). Numerical Models for Simulation the Fractional-Order Control Systems, *UEF-04-94, The Academy of Sciences, Inst. of Experimental Physic*, (Košice, Slovakia).

Dorčák, Ľ., Petráš, I., Koštial, I., and Terpák, J. (2002). Fractional-order state space models, in *Proc. of the International Carpathian Control Conference*, Malenovice, Czech rep, May 27-30, pp. 193–198.

Dorf, R. C., and Bishop, R. H. (1990). *Modern Control Systems* (Addison-Wesley, New York).

Ford, N., and Simpson, A. (2001). The numerical solution of fractional differential equations: speed versus accuracy, *Numerical Analysis Report 385*, (Manchester Centre for Computational Mathematics).

Gantmacher, F. R. (1959). *The Theory of Matrices*, (New York: Chelsea).

Gao, X., and Yu, J. (2005). Chaos in the fractional order periodically forced complex Duffing's oscillators, *Chaos, Solitons & Fractals*, vol. 24, pp. 1097–1104.

Ghartemani, M. K., and Bayat, F. M. (2008). Necessary and sufficient conditions for perfect command following and disturbance rejection in fractional order systems, in *Proc. of the 17th World Congress IFAC*, Soul, Korea, July 6-11, pp. 364–369.

Gorenflo, R., Luchko, Yu., and Rogosin, S. (2004). Mittag-Leffler type functions: notes on growth properties and distribution of zeros, *Preprint No. A-97-04, Fachbereich Mathematik und Informatik, Freie Universitt* (Berlin Germany).

Guo, L. J. (2005). Chaotic dynamics and synchronization of fractional-order Genesio-Tesi systems, *Chinese Physics*, vol. 14, no. 8, pp. 1517–1521.

Hao, B. (1989). *Elementary Symbolic Dynamics and Chaos in Dissipative Systems*, (Singapore: World Scientific).

Hartley, T. T., Lorenzo, C. F., and Qammer, H. K. (1995). Chaos on a fractional Chua's system, *IEEE Transactions on Circuits and Systems. Theory and Applications*, vol. 42, no. 8, pp. 485–490.

Hodges, D. A., Gray, P. R., and Broderson, R. W. (1978). Potenzial of MOS technologies for analog integrated circuits, *IEEE Journal of Solid-State Circuits*, **13** 3, pp. 285–294.

Hwang, Ch., Leu, J. F., and Tsay S. Y. (2002). A note on time-domain simulation of feedback fractional-order systems, *IEEE Transactions on Automatic Control*, **47**, 4, pp. 625–631.

Kaplan, W. (1992) *Advanced Calculus* (Addison Wesley).

Karmarkar, J. S., and Siljak, D. D. (1970). Stability analysis of systems with time delay, in *Proc. IEE*, **117**, 7, pp. 1421-1424.

Kennedy, M. P. (1992). Robust OP AMP realization of Chua's circuit, *Frequenz*, vol. 46, no. 3-4, pp. 66–80.

Keshner, M. S. (1982). 1/f Noise, *Proceedings of the IEEE*, **70**, 53, pp. 212–218.

Kharitonov, V. L., and Zhabko, A. P. (1994). Robust stability of time-delay systems, *IEEE Transaction Automatic Control*, **39**, 12, pp. 2388–2397.

Kilbas, A. A., Srivastava, H. M., and Trujillo, J. J. (2006). *Theory and Applications of Fractional Differential Equations* (Elsevier).

Kim, K. J., and Shahinpoor, M. (2003). Ionic polymermetal composites: II, *Manufacturing techniques and Smart Material Structures*, **12**, pp. 65-79.

Korabel, N., Zaslavsky, G. M., and Tarasov, V. E. (2007). Coupled oscillators with power-law interaction and their fractional dynamics analogues, *Communications in Nonlinear Science and Numerical Simulation*, **12**, 8, pp. 1405–1417.

Le Mehauté, A. (1991). *Fractal Geometries* (CRC Press).

Lepage W.R. (1961). *Complex variables and the Laplace transform for engineers* (McGraw-Hill).

Li, Y., Chen, Y.Q., Podlubny, I., and Cao, Y. (2008). Mittag-Leffler stability of fractional order nonlinear dynamic system, in *Proc. of the 3rd IFAC Workshop on Fractional Differentiation and its Applications* (Ankara, Turkey).

Li, C., and Yan, J. (2007). The synchronization of three fractional differential systems, *Chaos, Solitons & Fractals*, vol. 32, no. 2, pp. 751–757.

Li, C., and Chen G. (2004). *Chaos and hyperchaos in the fractional-order Rossler equations, Physica A*, vol. 341, pp. 55–61.

Lorenzo, C. F., Wang, Y., Hartley, T. T., Carletta, J. E., and Veillette, R. J. (2008). A Fractional-order Model for Ultracapacitor Long-term Behavior, in *Proc. IFAC Workshop on Fractional Differentiation and its Applications*, (FDA'08) (Ankara, Turkey).

Machado T., Da Graca Marcos, M., and Duarte, F. (2008). Fractional dynamics in the trajectory control of redundant manipulators, *Communications in Nonlinear Science and Numerical Simulation*, vol. 13, no. 9, pp. 1836–1844.

Magin, R. (2006). *Fractional Calculus in Bioengineering* (Begell House Publishers).

Manabe, S. (1961). The Non-Integer Integral and its Application to Control Systems. *ETJ of Japan*, vol. 6, no. 3-4, pp. 83–87.

Manabe, S. (2002). A Suggestion of fractional-order controller for flexible spacecraft attitude control, *Nonlinear Dyn.*, vol. 29, pp. 251–268.

Mandelbrot, B. (1967). Some noises with $1/f$ spectrum, a bridge between direct current and white noise, *IEEE Transaction Informormation Theory*, **13**, 2, pp. 289–298.

Marquardt, D. W. (1963). An Algorithm for Least-Squares Estimation of Nonlinear Parameters, *Journal of the Society for Industrial and Applied Mathematics*, **11**, 2, pp. 431–441.

Matignon, D. (1998). Stability properties for generalized fractional differential systems, in *Proc. of Fractional Differential Systems: Models, Methods and Applications*, pp. 145–158.

Matignon, D., and D'Andrea-Novel. B. (1996). Some results on controllability and observability of finite-dimensional fractional differential systems, in *Proc. of Computational Engineering in Systems Applications*, (Lille, France), pp. 952–956.

Matignon D. (1996). Stability result on fractional differential equations with applications to control processing, in *Proc. IMACS-SMC Proceedings*, Lille, France, pp. 963-968.

Matsumoto, T. (1984). *A chaotic attractor from Chua's circuit, IEEE Trans. on Circuit and Systems*, vol. CAS-31, no. 12, pp. 1055–1058.

Web site of Microchip Technology corporation. PIC18F458 processor *documentation*, http://www.microchip.com/.

Monje, C. A., Vinagre, B. M., Feliu, V., and Chen, Y. Q. (2008). Tuning and auto-tuning of fractional order controllers for industry application, *Control Engineering Practice*, vol. 16, pp. 798–812.

Nemat-Nasser, S. (2002). Micromechanics of Actuation of Ionic Polymer-metal Composites, *Journal of Applied Physics*, **92**, pp. 2899–2915.

Nemat-Nasser, S., and Wu, Y. (2003). Comparative experimental study of ionic polymer-metal composites with different backbone ionomers and in various cation forms, *Journal Applied Physics*, **93**, 9, pp. 5255-5267.

Oldham, K. B., and Spanier, J. (2006). *The Fractional Calculus: Theory and Applications of Differentiation and Integration to Arbitrary Order* (Dover Books on Mathematics).

Onaral, B., and Schwan, H. P.(1982). Linear and non linear properties of platinum electrode polarization, Part I, Frequency dependence at very low frequencies, *Medical Biological Engeneering Computation*, **20**, pp. 299–306.

Oustaloup, A., Sabatier,J., Lanusse, P., Malti, R., Melchior, P., Moreau, X., and Moze, M. (2008). An overview of the CRONE approach in system analysis, modeling and identification, observation and control, in *Proc. of the 17th World Congress IFAC*, Soul, Korea, July 6-11, pp. 14254–14265.

Oustaloup, A. (1995). *La Derivation Non Entiere: Theorie, Synthese et Applications*, Paris (Hermes).

Oustaloup, A., and Bansard, M. (1993). First generation CRONE control, in *Proc. International Conference on Systems, Man and Cybernetics*, Oct. 17-20, vol. 2, pp. 130–135.

Oustaloup, A., Lanusse, P., and Mathieu, B. (1993). Second generation CRONE control, in *Proc. International Conference on Systems, Man and Cybernetics*, Oct. 17-20, vol. 2, pp. 136–142.

Oustaloup, A., Lanusse, P., and Mathieu, B. (1993). Third generation CRONE control, in *Proc. International Conference on Systems, Man and Cybernetics*, Oct. 17-20, vol. 2, pp. 149–155.

Oustaloup, A. (1983) *Systemes asservis lineares d'ordre fractionnaire* (Masson, Paris).

Petráš, I., Dorčák, Ľ., and Koštial. I. (1998). A comparison of the integer and the fractional order controller on the laboratory object, in *Proc. of the ICAMC'98/ASRTP'98 conference*, Tatranske Matlaire, pp. 451–454.

Petráš, I., and Dorčák, Ľ. (1999). The Frequency Method for Stability Investigation of Fractional Control Systems, *Journal of SACTA*, **1**, 1-2, pp. 75–85.

Petráš, I. (1999). The fractional-order controllers: methods for their synthesis and application, *J. of Electrical Engineering*, vol. 50, no. 9-10, pp. 284–288.

Petráš, I. (2000). Fractional Calculus in Control, PhD Thesis, Technical University of Kosice.

Petráš, I., Podlubny, I., O'Leary, P., Dorčák, Ľ., and Vinagre, B. M. (2002). *Analogue Realizations of Fractional Order Controllers* (FBERG, TU Kosice).

Petráš, I., Vinagre, B. M., Dorčák, Ľ., and Feliu, V. (2002). Fractional digital control of a heat solid: Experimental results, in *Proc. of the ICCC'02*, Malenovice, Czech Republic, May 27-30, pp. 365–370.

Petráš, I. (2003). Digital Fractional Order Differentiator/integrator - FIR type, http://www.mathworks.com/matlabcentral/fileexchange/3673.

Petráš, I. (2003). Digital Fractional Order Differentiator/integrator - IIR type, http://www.mathworks.com/matlabcentral/fileexchange/3672.

Petráš, I., and Grega, Š. (2003). Digital fractional order controllers realized by PIC microprocessor: experimental results, in *Proceedings of the ICCC2003*, High Tatras, Slovak Republic, pp. 873–876.

Petráš, I., Dorčák, Ľ., Podlubny, I., Terpák, J., and O'Leary, P. (2005). Implementation of fractional-order controllers on PLC B&R 2005, in *Proceedings of the ICCC2005*, Miskolc-Lillafured, Hungary, May 24-27, pp. 141–144.

Petráš, I., Chen, Y. Q., Vinagre, B. M., and Podlubny, I. (2005). Stability of linear time invariant systems with interval fractional orders and interval coefficients, in *Proc. of the International Conference on Computation Cybernetics*, Vienna, Austria, pp. 1-4.

Petráš, I. (2008). A note on the fractional-order Chua's system, *Chaos, Solitons & Fractals*, vol. 38, no. 1, pp. 140–147.

Petráš, I. (2009). Chaos in the fractional-order Volta's system: modeling and simulation. *Nonlinear Dyn.*, vol. 57, no. 1-2, pp. 157–170.

Petráš, I. (2009). Fractional-Order Feedback Control of a DC Motor, *J. of Electrical Engineering*, vol. 60, no. 3, pp. 117–128.

Podlubny, I. (1999). *Fractional Differential Equations* (Academic Press, San Diego).

Podlubny, I. (1999). Fractional order systems and $PI^\lambda D^\mu$ controler, in *Proc. IEEE Transaction on Automatic Control*, **44**, 2, pp. 208–214.

Podlubny, I., Petráš, I., Vinagre, B. M., OLeary, P., and Dorčák, L. (2002). Analogue Realization of Fractional Order Controller, *Nonlinear Dyn.*, **29**, 1-4, pp. 281–296.

Podlubny, I. (2002). Matrix approach to discrete fractional calculus, *Fractional Calculus and Applied Analysis*, f3, 4, pp. 359–386.

Podlubny, I., Chechkin, A., Škovránek, T., Chen, Y. Q., and Vinagre, B. M. (2009). Matrix approach to discrete fractional calculus II: Partial fractional differential equations, *Journal of Computational Physics*, **228**, 8, pp. 3137–3153.

Pontryagin, L. S. (1995). On the zeros of some elementary transcendental function, *American Mathematical Society Translation*, **2**, pp. 95-110.

Radwan, A. G., Soliman, A. M., Elwakil, A. S., and Sedeek, A. (2009). On the stability of linear systems with fractional-order elements, *Chaos, Solitons & Fractals*, vol. 40, no. 5, pp. 2317–2328.

Ross, B. (editor) (1975). *Fractional Calculus and its Applications* (Springer-Verlag, Berlin).

Sabatier, J., Agrawal, O. P., and Tenreiro Machado, J. A. (2007). *Advances in Fractional Calculus: Theoretical Developments and Applications in Physics and Engineering* (Springer).

Sabatier, J., Bertrand, N., Briat, O., and Vinassa, J. (2008). An ultracapacitor non-linear fractional model, in *Proc. IFAC Workshop on Fractional Differentiation and its Applications*, (FDA'08) (Ankara, Turkey).

Shahinpoor, M., Bar-Cohen, Y., Simpson, J. O., and Smith, J. (1998). Ionic Polymer-Metal Composites (IPMC) as Biomimetic Sensors, Actuators, and Artificial Muscle, *Int. J. Smart Materials and Structures*, **7**, pp. R15-R30.

Shahinpoor, M. (1999). Electro-Mechanics of iono-elastic beams as electrically controllable artificial muscle, in *Proc. SPIE 6th Annual International Symposium on Smart Structures and Materials*, (Newport Beach, CA).

Shahinpoor, M., and Kim, K. J. (2001). Ionic polymermetal composites: I. Fundamentals, *Smart Materials and Structures* **10**, pp. 819-833.

Sheu, L. J., Chen, H. K., Chen, J. H., Tam, L. M., Chen, W. Ch., Lin, K. T., and Kang, Y. (2008). Chaos in the Newton–Leipnik system with fractional order, *Chaos, Solitons & Fractals*, vol. 36, no. 1, pp. 98–103.

Tarasov, V. E., and Zaslavsky, G. M. (2006). Fractional dynamics of systems with long-range interaction, *Communications in Nonlinear Science and Numerical Simulation*, **11**, 8, pp. 885–888.

Tarasov, V. E., and Zaslavsky, G. M. (2007). Conservation laws and Hamilton's equations for systems with long-range interaction and memory, *Commun. Nonlinear. Sci. Numer. Simulat.*, **13**, 9, pp. 1860–1878.

Tavazoei, M. S., and Haeri, M. (2008). Chaotic attractors in incommensurate fractional order systems, *Physica D*, **327**, pp. 2628–2637.

Tavazoei, M. S, and Haeri, M. (2008). Limitations of frequency domain approximation for detecting chaos in fractional order systems, *Nonlinear Analysis*, **69**, pp. 1299–1320.

Tavazoei, M. S., and Haeri, M. (2007). Unreliability of frequency-domain approximation in recognising chaos in fractional-order systems, *IET Signal Proc.*, **1**, 4, pp. 171–181.

Tavazoei, M. S., and Haeri M. (2007). A necessary condition for double scroll attractor existence in fractional - order systems, *Physics Letters A*, **367**, pp. 102–113.

Tavazoei, M. S., and Haeri, M. (2009). A note on the stability of fractional order systems, *Mathematics and Computers in Simulation*, vol. 79, no. 5, pp. 1566–1576

Tustin, A., Allanson, J. T., Layton, J. M., and Jakeways, R. J. (1958). The Design of Systems for Automatic Control of the Position of Massive Objects, *The Proceedings of the Institution of Electrical Engineers*, 105C (1).

Valerio, D., and da Costa, S. J. (2006). Tuning-Rules for Fractional PID Controllers, in *Proc. IFAC Workshop on Fractional Differentiation and its Application* (FDA'06) (Porto, Portugal).

Vinagre, B. M., and Feliu, V. (2007). Optimal Fractional Controllers for Rational Order Systems: A Special Case of the Wiener-Hopf Spectral Factorization Method, *IEEE Transactions on Automatic Control*, **52**, 12, pp. 2385-2389.

Vinagre, B. M., Chen, Y. Q., Dou, H., and Monje, C. A. (2006). Robust Tuning Method for Fractional Order PI Controllers, in *Proc. IFAC Workshop on Fractional Differentiation and its Application* (FDA'06) (Porto, Portugal).

Vinagre, B. M., Monje, C. A., Feliu, V., and Chen, Y. Q. (2006). *On Auto-Tuning of Fractional Order $PI^\lambda D^\mu$ Controllers*, in *Proc. IFAC Workshop on Fractional Differentiation and its Application* (FDA'06) (Porto, Portugal).

Vinagre, B. M., Chen, Y. Q., and Petráš, I. (2003). Two direct Tustin discretization methods for fractional-order differentiator/integrator, *Journal of Franklin Institute*, **340**, pp. 349-362.

Vinagre, B. M., Podlubny, I., Hernández, A., and Feliu, V. (2000). Some approximations of fractional order operators used in control theory and applications, *Fractional Calculus & Applied Analysis*, vol. 3, no. 3, pp. 231–248.

Vinagre, B. M., Podlubny, I., Dorčák, Ľ., and Feliu, V. (2000). *On Fractional PID Controllers: A Frequency Domain Approach*, Proc. of the IFAC Workshop on Digital Control, Terrassa, Spain, pp. 53–55.

Vinagre, B. M., Petráš, I., Merchan, P., and Dorčák, Ľ. (2001). Two digital realizations of fractional controllers: Application to temperature control of a solid, *Proc. of the ECC'01*, Porto, Portugal, September 4-7, pp. 1764–1767.

Vinagre, B. M., Monje, C. A., Calderon, A. J., Chen, Y. Q., and Feliu, V. (2004). The fractional integrator as reference function, *Proc. of the First IFAC Symposium on Fractional Differentiation and its Applications*, Bordeaux, France, July 19-20.

Wang, J. C. (1987). Realization of generalized Warburg impedance with RC ladder networks and transmission lines, *Journal Electrochemical Society*, **134**, 8, pp. 1915–1940.

Westerlund, S. (2002). *Dead Matter Has Memory!* (Kalmar, Sweden: Causal Consulting).

Yang, C., and Liu, F. (2006). A computationally effective predictor-corrector method for simulating fractional order dynamical control system, *Australian and New Zealand Industrial and Applied Mathematics Journal*, **47**, pp. C168–C184.

Index

www.ingramcontent.com/pod-product-compliance
Lightning Source LLC
Chambersburg PA
CBHW050627190326
41458CB00008B/2168